T0275990

SpringerBriefs in Electrical and Computer Engineering

More information about this series at http://www.springer.com/series/10059

Yuhua Xu · Alagan Anpalagan

Game-theoretic Interference Coordination Approaches for Dynamic Spectrum Access

 Springer

Yuhua Xu
PLA University of Science and Technology
Nanjing
China

Alagan Anpalagan
Ryerson University
Toronto
Canada

ISSN 2191-8112   ISSN 2191-8120   (electronic)
SpringerBriefs in Electrical and Computer Engineering
ISBN 978-981-10-0022-5   ISBN 978-981-10-0024-9   (eBook)
DOI 10.1007/978-981-10-0024-9

Library of Congress Control Number: 2015953807

Springer Singapore Heidelberg New York Dordrecht London

Printed on acid-free paper

Springer Science+Business Media Singapore Pte Ltd. is part of Springer Science+Business Media
(www.springer.com)

Preface

Dynamic spectrum access (DSA) is an efficient and promising approach to solve the dilemma between spectrum shortage and waste, which is mainly caused by the traditional static and pre-allocated spectrum access policies. The users in DSA systems have the ability to sense the environment, learn from history information, and then adjust their decisions in a smart and dynamic manner. Owing to the intelligent spectrum decision manner and open transmission in wireless communications, interference coordination among the interactive users becomes the primary concern.

Game theory is a powerful tool to study the interactions among multiple autonomous decision-makers. However, since it is a branch of applied mathematic, some new challenges with regard to information constraints should be addressed when it is applied to interference coordination for DSA systems. The purpose of this book is to bridge game theory and practical interference mitigation approaches, by taking into account the incomplete and dynamic information constraints in wireless communication networks. It establishes a game-theoretic framework and presents the cutting-edge technologies for distributed interference coordination. With game-theoretic formulation and the designed distributed learning algorithms, it provides insights on the interactions among the multiple decision-makers and the converging stable states. Furthermore, some promising and novel interference models are presented. We believe that it contains valuable knowledge, useful methods and practical algorithms that can be considered in emerging 5G wireless communication networks.

Many individuals have helped shape this book with their effort and time. We would like to thank Jinglong Wang, Qihui Wu, Liang Shen, Zhiyong Du, Youming Sun, Yuanhui Zhang and Yiwei Xu for their insightful contributions to this book.

Finally, thanks to Wayne Hu and Ivy Gong for their valuable advice throughout the production of this book.

This work was supported by the National Science Foundation of China under Grant No. 61401508.

Nanjing, China   Yuhua Xu
Toronto, Canada   Alagan Anpalagan

Contents

Acronyms

BR	Best response
CA	Carrier aggregation
CCC	Common control channel
CDMA	Code division multiple access
CR	Cognitive radio
CSMA	Carrier sensing multiple access
DSA	Dynamic spectrum access
MAC	Multiple access control
NE	Nash equilibrium
NOC	Non-partially overlapping channel
OFDMA	Orthogonal frequency division multiple access
POC	Partially overlapping channel
QoE	Quality of service
SAP	Spatial adaptive play
SE	Stackelberg equilibrium
TDMA	Time division multiple access

Chapter 1
Introduction

1.1 Interference Coordination in Dynamic Spectrum Access

1.1.1 Preliminaries

With the dramatically increasing demand in mobile traffic, the dilemma between spectrum shortage and waste, which is mainly caused by the traditional static and pre-allocated spectrum access policy, has became a serious problem facing the wireless communication systems. To solve this problem, dynamic spectrum access (DSA) is an efficient and promising approach. In recent years, based on the development of cognitive radio (CR) technology [1], in which the users have the ability to sense the environment, learn from history information, and then adjust their decisions in a smart and dynamic manner, DSA has drawn great attentions and can be used in several application scenarios. For example, in the primary-secondary spectrum access systems, the secondary users opportunistically access the channels which are unoccupied by the primary users; in heterogeneous networks, the small cells access the channels according to the spatial and temporary traffic distribution; in LTE-U systems, the users also dynamically access the licensed and unlicensed channels.

In DSA systems, the users perform spectrum access in a distributed and autonomous manner; furthermore, due to the feature of open transmission in wireless communications, interference becomes the primary concern [2]. Here, the considered interference metric is generalized. Specifically, it mainly includes: (i) The traditional PHY-layer interference: the interference signal received from other transmitters. (ii) The MAC-layer interference: in both time division multiple access (TDMA) and carrier sensing multiple access (CSMA) schemes, a user cannot transmit when its neighboring users are transmitting. In this scenario, the interference is different from the traditional PHY-layer interference as it only cares about the number of interfering users but not about the received interference signal. To capture this effect, the MAC-layer interference can be defined and analyzed. (iii) More generalized interference: we can generalize the interactions among the users when their objectives are

© The Author(s) 2016
Y. Xu and A. Anpalagan, *Game-theoretic Interference Coordination*
Approaches for Dynamic Spectrum Access, SpringerBriefs in Electrical
and Computer Engineering, DOI 10.1007/978-981-10-0024-9_1

conflicting. In this book, we will analyze the above-mentioned interference metrics in different scenarios.

1.1.2 Challenges and Problems

The key task in DSA systems is to choose the appropriate channels to mitigate mutual interference among multiple users. To summarize, there are some new challenges and problems:

1. Developing efficient distributed optimization approaches. In methodology, centralized optimization approaches have the following limitations: (i) they may cause heavy communication overhead as global information of all the users is required, and (ii) as the network scales up, the computational complexity becomes huge and unacceptable. Thus, it is desirable to develop efficient distributed optimization interference mitigation approaches.
2. Addressing the combinatorial nature of DSA problems. Generally, the decision variables in DSA are discrete, i.e., choosing a channel from the available channel set. For such combinatorial optimization problems, the well-known convex optimization approaches cannot be applied.
3. Coping with the incomplete and dynamic information constraints in wireless environment. Information is key to decision [3]. Unfortunately, incomplete information, e.g., a user only has its individual action-payoff information and partial environment information, and dynamic information constraints, e.g., the channel state is time-varying, cause challenges and new problems.

1.2 Game-Theoretic Solutions for Interference Coordination

Game theory [4] is a branch of applied mathematics, which provides efficient distributed approaches for solving problems involving multiple interactive decision-makers. Although it was originally studied in economics, it has been extensively applied into several application scenarios, e.g., biology [5], social activities [6], and engineering [7]. Since the pioneer work on applying game theory in power control [8, 9], it has been regarded as an important optimization tool for wireless networks [10, 11]. Naturally, it is very suitable for solving wireless optimization problems which are directly related to economic events and activities, e.g., spectrum auction [12–14] and incentive mechanism [15]. More importantly, it can be applied to solve any other involving multiple interactive users in wireless optimization problems, e.g., distributed power control [16], self-organizing networking [17], multiple access control [18], and heterogeneous network selection [19].

1.2.1 Motivation of Applying Game Models

In this book, we focus on developing game-theoretic interference coordination approaches for DSA systems [21]. The motivation of using game models are summarized as follows:

1. Game models provide a good and promising framework for distributed optimization, as the players in game take actions distributively and autonomously. Furthermore, the interactions among the players can be well analyzed and addressed using game theory.
2. The combinatorial nature of DSA problems can be easily addressed. In game models, the players choose actions from their available action set, which is always discrete. In this sense, it is believed that game models are very suitable for solving combinatorial optimization problems.
3. The dynamic and incomplete constraints can be solved, via careful design of repeated play in game models. Through repeated play, useful information can be obtained to guide the players to take actions. As a result, incomplete and dynamic information constraints can be addressed.
4. Smart and intelligent decision can be achieved. It is expected that learning is the core of future wireless communications [20]. On one hand, the outcome of the game is predicable and hence the performance can be improved. On the other hand, through repeated play, the players can learn from the past information and the feedback from the environment, adjust their behaviors, and finally achieve desirable and stable outcomes.

1.2.2 A General Framework of Game-Theoretic Solutions

1.2.2.1 Basic Game Models

Generally, a noncooperative game is denoted as $G = \{\mathbb{N}, A_n, u_n\}$, where

- $\mathbb{N}$ is the set of players. The player set can be defined flexibly. For example, in the spectrum auction problems, players are the operators. In network management and planning, the base stations with their serving clients are the players. In wireless access problems, the mobile users are the players.
- A_n is the available action set of player n. In most scenarios, the available action is a single decision variable, e.g., channel, time, power. However, it can also be defined as a combination of multiple decision variables, e.g., joint channel selection and power control, joint relay selection and power control.
- u_n is the utility function of n. Denote $a_n \in A_n$ as the chosen action of player n, and a_{-n} as the chosen action profile of all players except n. Then, the utility function is generally expressed as $u_n(a_n, a_{-n})$. In some scenarios, the utility of a user is only affected by the actions of neighboring users and hence the utility is

then determined by $u_n(a_n, a_{J_n})$, where J_n is the neighboring users and a_{J_n} is their chosen actions.

In some application scenarios, the players may choose mixed strategies over their available action sets. Formally, the mixed strategy of player n is denoted as $\sigma_n(a_n), a_n \in A_n$, which corresponds to the probability of player n choosing action a_n. The mixed strategy action space of player n is then given by $\Sigma_n = \{\sigma_n : \sum_{a_n \in A_n} \sigma_n(a_n) = 1, 0 \le \sigma_n(a_n) \le 1\}$. Denote σ_{-n} as the mixed strategy profile of all the players except n, then the expected achievable utility function of player n is determined by $u_n(\sigma_n, \sigma_{-n}) = \sum_{a \in A} \left(\prod_{k \in \mathbb{N}} \sigma_k(a_k) \right) u_n(a)$.

Since all the players in non-cooperative games are selfish and rational, i.e., they all maximize their utilities, it is important to study the stable solutions of the game. In the following, some important definitions are presented.

Definition 1.1 An action profile $a^* = (a_1^*, \dots, a_{-n}^*)$ is a pure strategy Nash equilibrium (NE) if and only if no player can improve its utility by deviating the current chosen action unilaterally, i.e.,

$$u_n(a_n^*, a_{-n}^*) \ge u_n(a_n, a_{-n}^*), \forall n \in \mathcal{N}, \forall a_n \in \mathcal{A}_n, a_n \neq a_n^* \tag{1.1}$$

The concept of NE was first coined by John. Nash [22], who was awarded the 1994 Nobel Prize in Economics. It is the most important solution concept in game theory. Based on NE, some other useful concepts of equilibria, e.g., correlated equilibrium (CE) [23], evolutionary stable strategy (ESS) [24], and conjectural equilibrium [25], are also extensively studied and used.

1.2.2.2 A General Framework of Game-Theoretic Solutions

For developing game-theoretic solutions for optimization problems in wireless communications, a general framework is shown in Fig. 1.1. It is seen that there are two key steps [26]: (i) game design and formulation, and (ii) distributed learning.

1. **Game design and formulation**. One needs to first identify the player set and the corresponding available action set, and then define the utility functions of the players. Defining utility function is very important since it inherently determines the properties and achievable performance of the game-theoretic models. There are three featured principles for defining utility function in wireless communications: (i) making the stable states optimal or near-optimal, which is the ultimate purpose of optimization in wireless communications, (ii) addressing the inherent features of wireless communications, e.g., channel fading, time-varying traffic, and user mobility, and (iii) having clear physical meanings. That is, it should explicitly be related to the optimization metrics in wireless communications, e.g., achievable throughput, interference, delay, or energy-efficiency.
2. **Design of distributed learning**. In pure game theory, it is always assumed that the players can perfectly monitor the environment and the actions chosen by

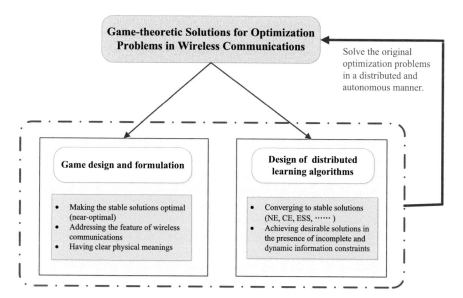

Fig. 1.1 The general framework of game-theoretic solutions for optimization problems in wireless communications

other players. As a result, some efficient algorithms, e.g., best response [27] and fictitious play [4], can be used to adjust their strategies toward stable solutions. However, in the presence of incomplete and dynamic information constraints in wireless communications, the task of achieving stable and desirable solutions is challenging. Thus, efforts should be given to: (i) developing efficient distributed learning algorithms which converge to stable solutions, e.g., NE, CE, and ESS, and (ii) achieving desirable stable solutions, e.g., maximizing the throughput or minimizing the interference.

Denote $a_n(k)$ as the action of player n in the kth iteration, and $a_{-n}(k)$ as the action profile of all other players except n. Due to the interactions (interference, congestion or competition) among the multiuser players, the received payoff $r_n(k)$ of each player is jointly determined by the action profile of all players, i.e., $r_n(k) = g_n(a_n(k), a_{-n}(k))$, where the payoff function $g_n(\cdot)$ may be determinate or random. Generally, the players perform the following learning procedure to update their actions:

$$a_n(k+1) = F(a_n(k), a_{-n}(k); r_n(k), r_{-n}(k)), \qquad (1.2)$$

Since the action update of a player is based on the profiles of chosen action and received payoff in the last iteration, the system evolution can be described as $\{a_n(k), a_{-n}(k)\} \to \{r_n(k), r_{-n}(k)\} \to \{a_n(k+1), a_{-n}(k+1)\}$. Thus, the optimization objective of the learning algorithm is to find a stable action profile to maximize the system utility.

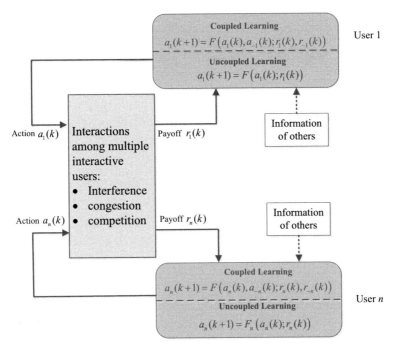

Fig. 1.2 The illustrative diagram of distributed learning algorithms in games

As discussed before, the incomplete and dynamic information constraints may pose some challenges. Specifically, (i) a player may not know the information about all other players, (ii) the received payoff $r_n(k)$ may be random and time-varying. Thus, the update rule needs to be carefully designed to guarantee the convergence toward desirable solutions. In addition, the update rule given in (1.2) is *coupled* since it needs to know information about others. To reduce the information exchange among users, it is desirable to develop *uncoupled* learning algorithms, i.e.,

$$a_n(k + 1) = F_u(a_n(k), r_n(k)), \qquad (1.3)$$

where only the individual information of action and received payoff are needed. An illustrative diagram of distributed learning in games is shown in Fig. 1.2.

1.3 Organization and Summary

In the following, we present the definition of exact potential game [27], which admits promising properties and have been extensively used in wireless communication networks [28]. Potential game admits several promising properties and the most

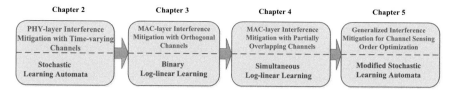

Fig. 1.3 Summary of game-theoretic interference coordination approaches in this book

important two are [27]: (i) every potential game has at least one pure strategy NE, (ii) the global or local maxima of the potential function correspond to a pure strategy NE. Furthermore, there are several efficient distributed learning algorithms which converges to NE of potential games in the presence of incomplete and dynamic information constraints.

Definition 1.2 A game is a potential game (EPG) if there exists an exact potential function $\phi_e : A_1 \times \cdots \times A_N \to R$ such that for all $n \in \mathbb{N}$, all $a_n \in A_n$, and $a'_n \in A_n$,

$$u_n(a_n, a_{-n}) - u_n(a'_n, a_{-n}) = \phi_e(a_n, a_{-n}) - \phi_e(a'_n, a_{-n}) \qquad (1.4)$$

That is, the change in the individual utility function caused by the unilateral action change of an arbitrary user is the same with that in the exact potential function. In this book, several potential game-based interference coordination approaches are presented and analyzed, the summary of game-theoretic interference coordination approaches in this work is shown Fig. 1.3. Specifically, the rest of this book is organized as follows.

- In Chap. 2, an interference mitigation game with time-varying channels is formulated and a stochastic learning automata-based algorithm is applied. The distinct feature is that the channel gains are randomly changing, which is common in practical wireless networks.
- In Chap. 3, an MAC-layer interference mitigation game with orthogonal channels is formulated and the binary log-linear learning algorithm is applied.
- In Chap. 4, an MAC-layer interference mitigation game with partially overlapping channels is formulated and the simultaneous log-linear learning algorithm is applied.
- In Chap. 5, a generalized interference mitigation game for sequential channel sensing and access is formulated and a modified stochastic learning automata-based algorithm is applied. The distinct feature is that the active user set is time-varying.
- Finally, future direction and research issues are presented in Chap. 6.

References

1. S. Haykin, Cognitive radio: brain-empowered wireless communications. IEEE J. Sel. Areas Commun. **23**(2), 201–220 (2005)
2. D. Gesbert, S.G. Kiani, A. Gjendemsjø et al., Adaptation, coordination, and distributed resource allocation in interference-limited wireless networks. Proc. IEEE **95**(12), 2393–2409 (2007)
3. S. Haykin, Me Fatemi, P. Setoodeh, Cognitive control. Proc. IEEE **100**(12), 3156–3169 (2012)
4. R. Myerson, *Game Theory: Analysis of Conflict* (Harvard University Press, Cambridge, 1991)
5. R. Axelrod, W.D. Hamilton, The evolution of cooperation. Science **211**(4489), 1390–1396 (1981)
6. E. Fehr, U. Fischbacher, The nature of human altruism. Nature **425**(6960), 785–791 (2003)
7. C.S. Yeung, A.S. Poon, F.F. Wu, Game theoretical multi-agent modelling of coalition formation for multilateral trades. IEEE Trans. Power Syst. **14**(3), 929–934 (1999)
8. A.B. MacKenzie, S.B. Wicker, Game theory and the design of self-configuring, adaptive wireless networks. IEEE Commun. Mag. **39**(11), 126–131 (2001)
9. H. Yaïche, R.R. Mazumdar, C. Rosenberg, A game theoretic framework for bandwidth allocation and pricing in broadband networks. IEEE/ACM Trans. Netw. (TON) **8**(5), 667–678 (2000)
10. Z. Han, D. Niyato, W. Saad, T. Basar, A. Hjorungnes, *Game Theory in Wireless and Communication Networks* (Cambridge University Press, Cambridge, 2012)
11. H. Tembine, *Distributed Strategic Learning for Wireless Engineers* (CRC Press, Boca Raton, 2012)
12. D. Niyato, E. Hossain, Z. Han, Dynamics of multiple-seller and multiple-buyer spectrum trading in cognitive radio networks: a game-theoretic modeling approach. IEEE Trans. Mob. Comput. **8**(8), 1009–1022 (2009)
13. L. Gao, X. Wang, Y. Xu, Q. Zhang, Spectrum trading in cognitive radio networks: a contract-theoretic modeling approach. IEEE J. Sel. Areas Commun. **29**(4), 843–855 (2011)
14. D. Niyato, E. Hossain, Spectrum trading in cognitive radio networks: a market-equilibrium-based approach. IEEE Wirel. Commun. **15**(6), 71–80 (2008)
15. D. Yang, G. Xue, X. Fang, and J. Tang, Crowdsourcing to smartphones: incentive mechanism design for mobile phone sensing, in *Proceedings of the 18th Annual International Conference on Mobile Computing and Networking* (2012), pp. 173–184
16. C.U. Saraydar, N.B. Mandayam, D. Goodman, Efficient power control via pricing in wireless data networks. IEEE Trans. Commun. **50**(2), 291–303 (2002)
17. V. Srivastava, J.O. Neel, A.B. MacKenzie, R. Menon, L.A. DaSilva, J.E. Hicks, J.H. Reed, R.P. Gilles, Using game theory to analyze wireless ad hoc networks. IEEE Commun. Surv. Tutor. **7**(1–4), 46–56 (2005)
18. K. Akkarajitsakul, E. Hossain, D. Niyato, D.I. Kim, Game theoretic approaches for multiple access in wireless networks: a survey. IEEE Commun. Surv. Tutor. **13**(3), 372–395 (2011)
19. R. Trestian, O. Ormond, G.-M. Muntean, Game theory-based network selection: solutions and challenges. IEEE Commun. Surv. Tutor. **14**(4), 1212–1231 (2012)
20. A. He, K.K. Bae, T.R. Newman et al., A survey of artificial intelligence for cognitive radios. IEEE Trans. Vech. Tech. **59**(4), 1578–1592 (2010)
21. Y. Xu, A. Anpalagan, Q. Wu et al., Decision-theoretic distributed channel selection for opportunistic spectrum access: strategies, challenges and solutions. IEEE Commun. Surv. Tutor. **15**(4), 1689–1713 (2013)
22. J. Nash, Non-cooperative games, Ann. Math. 286–295 (1951)
23. R.J. Aumann, Correlated equilibrium as an expression of Bayesian rationality, Econ.: J. Econ. Soc. 1–18 (1987)
24. P.D. Taylor, L.B. Jonker, Evolutionary stable strategies and game dynamics. Math. Biosci. **40**(1), 145–156 (1978)
25. M.P. Wellman, J. Hu, Conjectural equilibrium in multiagent learning. Mach. Learn. **33**(2–3), 179–200 (1998)

26. Y. Xu, J. Wang, Q. Wu, Z. Du, L. Shen, A. Anpalagan, A game theoretic perspective on self-organizing optimization for cognitive small cells. IEEE Commun. Mag. **53**(7), 100–108 (2015)
27. D. Monderer, L.S. Shapley, Potential games. Games Econ. Behav. **14**, 124–143 (1996)
28. K. Yamamoto, A comprehensive survey of potential game approaches to wireless networks. IEICE Trans. Commun. **E98–B**(9) (2015)

Chapter 2
Distributed Interference Mitigation in Time-Varying Radio Environment

2.1 Introduction

Currently, most existing studies on the problem of interference mitigation, e.g., [1–10], have assumed that the interference channel gains are static. Based on such an ideal assumption, there are several nongame theoretic [1, 9] and game-theoretic [2–8, 10] interference mitigation approaches. However, the assumption of static channels is not true since they are always time-varying in practice, which is the inherent feature of wireless communications.

In this chapter, we consider a multiuser, multichannel opportunistic spectrum access network, where the users choose orthogonal channels to mitigate mutual interference [4, 5, 7–10]. The considered network is completed distributed, as there is no centralized controller and no information exchange among users. To address the time-varying nature of wireless communication, it is assumed that the channels undergo block-fading. Block-fading means that the channel gains remain unchanged in a slot but change randomly in the next slot, which is realistic and has been extensively used in the past literature.

Following the similar ideas proposed in [6, 9, 10], in which the weighted aggregate interference for static channels is minimized, the network utility in this chapter is naturally extended to the expected weighted aggregate interference for time-varying channels. As a result, the optimization objective is to find channel selection profiles that minimize this network utility in a distributed manner. Since the channel selections of the users are distributed and autonomous, we formulate the problem of opportunistic spectrum access as a noncooperative game. With the formulated game model, we then propose a stochastic automata-based distributed learning algorithm, which converges to pure strategy NE of the interference mitigation game in time-varying environment. Note that the main analysis and results in this chapter were presented in [11].

© The Author(s) 2016
Y. Xu and A. Anpalagan, *Game-theoretic Interference Coordination Approaches for Dynamic Spectrum Access*, SpringerBriefs in Electrical and Computer Engineering, DOI 10.1007/978-981-10-0024-9_2

2.2 System Model and Problem Formulation

2.2.1 System Model

We consider a distributed canonical wireless network consisting of multiple autonomous users. Note that each user in canonical networks is not a single communication entity but a collection of multiple entities with intracommunications [12–14]. Generally, there is a leading entity choosing the operational channel and the belonged members share the channel using some multiple access control mechanisms, e.g., TDMA or CSMA/CA. Examples of wireless canonical network are given by, e.g., a WLAN access point with the serving clients [12] and a cluster head together with its members [9]. A comprehensive review on canonical networks can be found in [9]. An illustrative example of the considered canonical networks is shown in Fig. 2.1.

Suppose that there are N users and M channels, and each user chooses one channel for communication. Denote the user set as $\mathbb{N} = \{1, \ldots, N\}$ and the channel set as $\mathbb{M} = \{1, \ldots, M\}$. To capture the time-variation of channels, it is assumed that all the channels undergo block-fading, i.e., the channel gains are block-fixed in a time slot and change randomly in the next slot. Furthermore, each user chooses exactly one channel for intra-communication at a time. When two users, say m and n, choose a channel simultaneously, mutual interference emerges, the instantaneous interference gain from users m to n in a specific slot can be expressed as:

$$w_{mn}^s = (d_{mn})^{-\alpha} \varepsilon_{mn}^s, \tag{2.1}$$

where the superscript s denotes the selected channel, d_{mn} is the physical distance between m and n, α is the path loss exponent, and ε_{mn}^s is the instantaneous random component of the path loss [15], e.g., Rayleigh fading. Due to fading in wireless environment, the instantaneous random components between two users in each slot are generally different. However, their expected values are assumed to be the same. Therefore, we can denote the expected value of the random components between any two users on a channel as $\bar{\varepsilon}_{mn}^s = \mathbf{E}[\varepsilon_{mn}^s] = \mathbf{E}[\varepsilon_{nm}^s], \forall m, n \in \mathbb{N}, \forall s \in \mathbb{M}$.

Fig. 2.1 An illustrative example of canonical networks

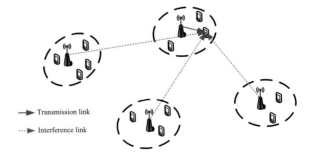

Remark 2.1 The interference channel model characterized by (5.2) is very general, since the instantaneous random components ε_{mn}^s can vary from slot to slot, from channel to channel, and from user to user. Furthermore, the dynamics may be independent or correlated. In addition, the expected value of random component $\bar{\varepsilon}_{mn}^s$ can also vary from channel to channel and from user to user. Thus, the analysis and results obtained in this chapter suitable for several practical scenarios, and some examples are given by: (i) when it is unit-constant, i.e., $\varepsilon_{mn}^s = 1, \forall m, n, s$, it corresponds to a scenario where only large-scale power-loss is considered, (ii) when it is log-normal distribution, it corresponds to the medium-scale power-loss, and (iii) when it is Rayleigh/Nakagami distribution, which means that multiple-path power-loss is considered.

2.2.2 Problem Formulation

Denote the chosen channel of user n in a slot as a_n, $a_n \in \mathbb{M}$, then the instantaneous achievable rate of user n is given by:

$$R_n = B \log \left(1 + \frac{p_n w_{nn}^{a_n}}{BN_0 + I_n}\right), \tag{2.2}$$

where B is the channel bandwidth, $w_{nn}^{a_n} = (d_{nn})^{-\alpha}\varepsilon_{nn}^{a_n}$ is the intracommunication channel gain of user n (the channel gain between the head and the serving clients), p_n is the transmitting power, N_0 is the noise power spectrum density, and I_n is the aggregate interference experienced by user n. For an action selection profile of all the users $a = \{a_1, \ldots, a_N\}$, I_n is random and can be expressed by:

$$I_n = \sum_{m \in \{\mathbb{N}\backslash n\}} f(a_m, a_n) p_m w_{mn}^{a_n}, \tag{2.3}$$

where $X \backslash Y$ means that Y is excluded from the set X, and $f(\cdot)$ is the following indicator function:

$$f(x, y) = \begin{cases} 1, & x = y \\ 0, & x \neq y. \end{cases} \tag{2.4}$$

According to (2.2), the aggregate expected network rate achieved by all the users can be expressed as:

$$R^{\text{sum}} = \sum_{n \in \mathbb{N}} \mathbf{E}[R_n] \tag{2.5}$$

From the perspective of interference mitigation, we consider the expected weighted aggregate interference in the network, which is defined as:

$$U = \sum_{n \in \mathbb{N}} p_n \mathbf{E}[I_n] = \sum_{n \in \mathbb{N}} \sum_{m \in \{\mathbb{N} \setminus n\}} p_m p_n \bar{w}_{mn}^{a_n} f(a_m, a_n), \qquad (2.6)$$

where $\bar{w}_{mn}^{a_n} = \mathbf{E}[w_{mn}^{a_n}] = (d_{mn})^{-\alpha} \bar{\varepsilon}_{mn}^{a_n}$ is the expected interference gain from user m to user n in channel a_n.

Note that the considered network utility metric, i.e., the weighted aggregate interference, has been studied in previous studies [6, 9, 10]. In [6], it was shown that such a network utility can balance the transmitting power and the experienced interference. Furthermore, it has been shown that with this network utility, near-optimal network rate can be achieved in low SINR regime [9]. Existing studies were mainly for static scenarios with fixed channel gains. In comparison, in order to address the random and instantaneous fading components, i.e., ε_{mn}^s, in wireless environment, we consider the expected version of weighted aggregate interference here. Therefore, motivated by the previous researches on interference mitigation rather than maximizing throughput directly, e.g., [4, 5, 9], the considered objective here is to minimize the expected weighted aggregate interference, as specified by (2.6), i.e.,

$$(P1 :) \quad a_{opt} \in \arg \min_a U \qquad (2.7)$$

2.3 Interference Mitigation Game in Time-Varying Environment

As the decision variable (channel selection) is discrete, the interference mitigation problem $P1$ is a combinatorial optimization problem. On the condition that all the key parameters including p_n, d_{mn} and $\bar{\varepsilon}_{mn}^s$, $\forall m, n \in \mathbb{N}, s \in \mathbb{M}$ are a priori known, centralized approaches can be applied. However, if there is no centralized control and these parameters are unknown, which is exactly the scenario considered in this chapter, the task of solving $P1$ is challenging. In the following, we propose a game-theoretic distributed approach in time-varying environment.

2.3.1 Game Model

The problem of distributed channel selection for interference mitigation in canonical networks is formulated as a noncooperative game. Formally, the game is denoted as $G_c = [\mathbb{N}, \{A_n\}_{n \in \mathbb{N}}, \{u_n\}_{n \in \mathbb{N}}]$, where $\mathbb{N} = \{1, \dots, N\}$ is the player set, $A_n = \{1, \dots, M\}$ is the available actions (channel) set for each player n, and u_n is the utility function of player n. As the experienced interference is a random variable in each slot, we consider the following utility function, which is defined as the expected experienced interference, i.e.,

$$u_n(a_n, a_{-n}) = D - p_n \mathbf{E}[I_n] = D - \sum_{m \in \{\mathbb{N} \backslash n\}} p_n p_m \bar{w}_{mn}^{a_n} f(a_m, a_n), \tag{2.8}$$

where a_{-n} is the channel selection profile of all the players except player n, I_n is the experienced interference of player n, as specified by (2.3), and D is a predefined positive constant which will be illustrated later. Then, the proposed interference mitigation game is expressed as:

$$(G): \quad \max_{a_n \in A_n} u_n(a_n, a_{-n}), \quad \forall n \in \mathbb{N}. \tag{2.9}$$

2.3.2 Analysis of Nash Equilibrium

In the following, we analyze the Nash equilibrium (NE) of the formulated interference mitigation game and investigate its properties.

Theorem 2.1 *The formulated interference mitigation game G_c is an exact potential game which has at least a pure strategy NE point, and the optimal channel selection that globally minimizes the expected weighted aggregate interference constitutes a pure strategy NE point of G.*

Proof Detailed lines for the proof are omitted here but can be found in [11]. In the following, only the proof skeleton is presented. First, we construct the following potential function:

$$\Phi(a_n, a_{-n}) = -\frac{1}{2} \sum_{n \in \mathbb{N}} \sum_{m \in \{\mathbb{N} \backslash n\}} p_m p_n \bar{w}_{mn}^{a_n} f(a_m, a_n), \tag{2.10}$$

which immediately yields the following equation:

$$\Phi(a_n, a_{-n}) = -\frac{1}{2} U(a_n, a_{-n}), \tag{2.11}$$

through which the network utility $U(a_n, a_{-n})$, as specified by (2.6), is related to the potential function. Then, after some mathematical manipulations, it can be verified that the change in individual utility function caused by any player's unilateral deviation is the same as that in the potential function. Thus, according to the definition given in Chap. 1, it is known that G is an exact potential game with Φ serving as the potential function. Therefore, Theorem 5.1 is proved. □

Theorem 5.1 characterizes the relationship between the interference mitigation game G and the network utility in general network scenarios. For further investigation, the following three scenarios are considered [3]: (i) under-loaded scenario: the number of users is less than that of channels, i.e., $N < M$, (ii) equally-loaded scenario: the number of users is equal to that of channels, i.e., $N = M$, and (iii) over-loaded scenario: the number of users is greater than that of channels, i.e., $N > M$. Then, the properties for the three scenarios are characterized by the following propositions, respectively.

Proposition 2.1 *For both under-loaded or equally-loaded scenarios, any pure strategy NE of the interference mitigation game G leads to an interference-free channel selection profile.*

Proof In the two scenarios, all pure strategy NE points correspond to orthogonal channel selection profiles, i.e., a channel is selected by no more than one user. This argument is due to the fact that no user is willing to deviate, as it experiences zero interference. Therefore, any pure strategy NE point is optimal to $P1$, and makes the network interference-free. Therefore, Proposition 2.1 is proved. □

Proposition 2.2 *For the over-loaded scenario, there exists at least one pure strategy NE point that minimizes the expected weighted aggregate interference.*

Proof Multiple pure strategy NE points may exist in the over-loaded scenario but the number of pure strategy NE is hard to obtain. However, according to Theorem 5.1, there is at least one pure strategy NE minimizing the expected weighted aggregate interference. Besides the optimal one, other pure strategy NE points only locally minimize the expected weighted aggregate interference. □

Since the global optimality is not guaranteed in the over-loaded scenarios, it is indispensable to study the performance of NE solutions. Generally, the concept of price of anarchy (PoA) [16] is used to study the performance ratio between the worst NE solution and the social optimum. However, as the PoA for the formulated game is hard to derive, we get an upper bound instead. To begin with, the achievable expected aggregate interference at a pure strategy NE $a^* = (a_1^*, \ldots, a_N^*)$ is given by:

$$U_{NE} = \sum_{n \in \mathbb{N}} p_n \mathbf{E}[I_n] = \sum_{n \in \mathbb{N}} \sum_{m \in \{\mathbb{N} \setminus n\}} p_m p_n \bar{w}_{mn}^{a_n^*} f(a_m^*, a_n^*). \tag{2.12}$$

Proposition 2.3 *If the values of the expected random components of all channels are the same, i.e., $\bar{\varepsilon}_{mn}^s = \bar{\varepsilon}_{mn}^0$, $\forall m, n \in \mathbb{N}$, then the expected aggregate interference of any pure strategy NE solution in an over-loaded scenario is upper bounded by $U_{NE} \leq U_0/M$, where*

$$U_0 = \sum_{n \in \mathbb{N}} \sum_{m \in \{\mathbb{N} \setminus n\}} p_n p_m (d_{mn})^{-\alpha} \bar{\varepsilon}_{mn}^0 \tag{2.13}$$

can be regarded as the expected aggregate interference if all the players choose the same channel.

Proof Refer to [11]. □

Remark 2.2 Generally, U_0 is the worst-case of the expected aggregate interference of an arbitrary network. According to Proposition 2.3, we can see that increasing the number of channels, i.e., M, would decrease the aggregate interference in the network, which can be expected in any wireless networks.

2.4 Achieving NE Using Stochastic Learning Automata

With the interference mitigation problem formulated as a potential game, the next task is to develop distributed learning algorithm to achieve NE. Notably, we encounter with the following incomplete and dynamic information constraints: (i) obtaining information of other players is not feasible, and (ii) the interference channel gains vary randomly from slot to slot. As a result, the commonly used learning algorithms for potential games, e.g., best response dynamic [17], no-regret learning [4], fictitious play [18], and spatial adaptive play [19], cannot be applied. To overcome this problem, we propose a stochastic learning automata [20]-based algorithm, which is simple and completely distributed.

2.4.1 Algorithm Description

To begin with, the game is extended to a mixed strategy form. Specifically, the mixed strategy for player n at iteration k is denoted by the probability distribution $q_n(k) \in \Delta(A_n)$, where $\Delta(A_n)$ is the set of all possible probability distributions over the action set A_n. In the stochastic learning automata algorithm, the game is played only once in a slot. After each play, each player receives a random payoff, which is jointly determined by action profiles of all the users and the instantaneous channel gains. Based on the received payoffs, the players update their mixed strategies using a simple and distributed rule. An illustrative diagram of the stochastic learning automata-based algorithm is shown in Fig. 2.2.

Suppose that at the kth slot, the channel selection profile of the users is $a(k) = \{a_1(k), \ldots, a_N(k)\}$. Then, the random payoff received by player n is as follows:

$$r_n(k) = D - \sum_{m \in \{\mathbb{N} \setminus \{n\}\}} p_m p_n (d_{mn})^{-\alpha} \varepsilon_{mn}^{a_n(k)} f(a_m(k), a_n(k)), \qquad (2.14)$$

where $f(\cdot)$ is the indicator function specified by (3.8), and $\varepsilon_{mn}^{a_n(k)}$ is the instantaneous channel gain. The purpose of adding the predefined positive constant D to the payoff, is to keep it positive. However, the received payoff may also be negative due to the fluctuation of random channel fading. Thus, the following modified received payoff is used in the distributed learning algorithm:

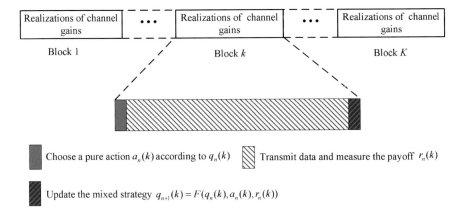

| Realizations of channel gains | ... | Realizations of channel gains | ... | Realizations of channel gains |

Block 1   Block k Block K

Choose a pure action $a_n(k)$ according to $q_n(k)$ Transmit data and measure the payoff $r_n(k)$

Update the mixed strategy $q_{n+1}(k) = F(q_n(k), a_n(k), r_n(k))$

Fig. 2.2 The schematic diagram of the stochastic automata learning-based channel selection algorithm

---

Algorithm 1: the stochastic learning automata-based channel selection algorithm

---

Initialization: set $k = 0$ and the mixed strategy of each user as $q_{ns}(k) = 1/|A_n|, \forall n \in \mathbb{N}, \forall s \in \mathbb{M}$.
Loop for $k = 1, 2, \ldots,$
1. Accessing the channels: In the kth slot, user n access a channel $a_n(k)$ according to its current selection probability vector $q_n(k)$.
2. Measuring payoffs: The game is played once with the channel selection profile $\{a_1(k), \cdots, a_N(k)\}$, and then all the players measure the received payoffs $r_n(k)$ using (2.15). Note that the payoff is random but can be directly measured by each user n [9].
3. Updating the mixed strategies: All the players update their mixed strategies using the following rules:

$$\begin{aligned} q_{ns}(k+1) &= q_{ns}(k) + b\tilde{r}_n(k)(1 - q_{ns}(k)), \ s = a_n(k) \\ q_{ns}(k+1) &= q_{ns}(k) - b\tilde{r}_n(k)q_{ns}(k), \quad\quad s \neq a_n(k), \end{aligned} \tag{2.16}$$

where $0 < b < 1$ is the learning step size, $\hat{r}_n(k)$ is the normalized received payoff which is given by:

$$\tilde{r}_n(k) = r_n(k)/D. \tag{2.17}$$

End loop

$$r_n(k) = \max\{r_n(k), 0\} \tag{2.15}$$

The stochastic learning automata-based algorithm is described in Algorithm 1. It is noted that the algorithm is online and fully distributed, as the users adjust the channel selections from their action-payoff history.

The proposed stochastic learning automata-based algorithm is also called linear reward-inaction (L_{R-I}), which is a special case of linear learning automata [20]. The updating rules for linear learning automata are generally given by:

$$q_n(k+1) = q_n(k) + bF\big(q_n(k), a_n(k), r_n(k)\big), \tag{2.18}$$

where $F(\cdot, \cdot, \cdot)$ is a learning function that maps the current action and payoff to the mixed strategy in the next iteration. Of course, other forms of update rules, e.g., linear reward-penalty and linear reward-ε-penalty [20], can also be used. The reason of using L_{R-I} is that it is simple and can be analyzed when being incorporated with game theory, which will be discussed below. Also, it is noted from (2.18) that it is only relying on the individual trial-payoff history of a player and does not need to know any information of others. In fact, each user is not even aware of other users.

2.4.2 Convergence Analysis

Using the method of stochastic approximation [21], the long-term behavior of the mixed strategies of the users can be characterized by an ordinary differential equation. Specifically, the convergence of the stochastic learning automata algorithm is characterized by the following theorem.

Theorem 2.2 *With a sufficiently small step size b, the stochastic learning automata-based learning algorithm asymptotically converges to a pure strategy NE point of an exact potential game.*

Proof Refer to Theorem 5 in [22]. □

Based on Theorem 2.2, the aggregate interference performance of the proposed game-theoretic interference mitigation solutions are characterized by the following propositions.

Proposition 2.4 *In under-loaded or equally-loaded scenarios, the proposed game-theoretic solution asymptotically converges to an optimal channel selection profile that makes the network interference-free.*

Proof This proposition can be proved by straightforwardly combining Theorem 2.2 and Proposition 2.1. □

Proposition 2.5 *In an over-loaded scenario, the proposed game-theoretic solution asymptotically converges a pure strategy channel selection profile and minimizes the expected weighted aggregate interference globally or locally.*

Proof According to Proposition 2.2, there is at least an optimal channel selection minimizing the aggregate interference, and they may be other suboptimal solutions. Thus, Proposition 2.5 is proved. □

Since there are various fading models, e.g., Rayleigh, Nakagami, and log-normal, it is important to study the achievable performance for different fading models. The following proposition reveals an interesting result.

Proposition 2.6 *For a given distributed network, the achievable interference performance of the proposed game-theoretic solution is determined by the expected interference gain but not the specific fading model.*

Proof Based on (2.6), it is seen that the expected weighted aggregate interference is jointly determined by user locations, the transmitting power, the final channel selection profile, and the expected interference gain $\bar{\varepsilon}_{mn}^s$. Thus, for a given distributed network, the achievable performance is only determined by the expected interference gain but the specific fading model.   □

According to Proposition 2.6, two fading models with the same expected fading gain, e.g., Rayleigh and Nakagami, would lead to the same expected weighted aggregate interference. Moreover, for a given fading model with unit-mean, the resulting expected weighted aggregate interference would be equal to a nonfading scenario, where only large-scale power-loss is considered.

The above analysis is for time-varying radio environment. As the static environment is an extreme case of time-varying case, we can conclude that stochastic learning automata-based algorithm also converges in static environment.

Proposition 2.7 *In a static system with symmetrical interference channels, the proposed game-theoretic solution also asymptotically converges to a pure strategy NE point of the channel selection game.*

Proof The experienced interference of a user in a static system is expressed as:

$$\hat{I}_n = \sum_{m \in \{\mathbb{N} \setminus n\}} f(a_m, a_n) p_m \hat{w}_{mn}^{a_n}, \tag{2.19}$$

where $\hat{w}_{mn}^{a_n}$ is the fixed interference gain from users m to n on channel a_n satisfying $\hat{w}_{mn}^{a_n} = \hat{w}_{nm}^{a_n}$. Then the aggregate weighted interference in a static network is given by:

$$\hat{U} = \sum_{n \in \mathbb{N}} p_n \hat{I}_n = \sum_{n \in \mathbb{N}} \sum_{m \in \{\mathbb{N} \setminus n\}} p_m p_n \hat{w}_{mn}^{a_n} f(a_m, a_n). \tag{2.20}$$

Similarly, a static channel selection game G_c with the following utility function can be defined:

$$\hat{u}_n(a_n, a_{-n}) = D - p_n \hat{I}_n. \tag{2.21}$$

Using similar lines of proof for Theorem 5.1, it can be proved that the channel selection game in static environment is also a potential game with potential function $-\frac{1}{2}\hat{U}$. Based on this result, we can prove this proposition following the same methodology in Theorem 2.2.   □

2.5 Simulation Results and Discussion

The simulation setting is similar to [9], in which the users are randomly located in a 100 m × 100 m region. For presentation, the transmitting powers of all the users are set assumed to be $p_n = 0\,\mathrm{dBw}, \forall n \in \mathbb{N}$, the path loss exponent is $\alpha = 2$, and the noise power as $N_0 = -130\,\text{dBw}$. For simplicity of analysis, the distance between the transmitter and the receiver for each intracommunication is set to 1 m, i.e., $d_{nn} = 1, \forall n \in \mathbb{N}$; The channel bandwidth is 1 MHz. We consider three common fading models: Rayleigh, Nakagami, and log-normal.

2.5.1 Convergence Behavior

2.5.1.1 Convergence Behavior in Dynamic Environment

In this part, we investigate the convergence with time-varying channel gains. Specifically, we consider a network with three channels and five users. Rayleigh fading with unit mean is considered. The positive constant used in the instantaneous received payoff (5.8) and (2.14) is set to $D = 0.005$, and the step size of the learning algorithm is set to $b = 0.1$.

The convergence behavior of three arbitrarily selected users is shown in Fig. 2.3. Taking user 1 as an illustrative example, it chooses the channels with equal probabilities at the beginning ($q_{11} = 0.33, q_{12} = 0.33, q_{13} = 0.33$), and finally chooses channel 3 ($q_{11} = 1, q_{12} = 0, q_{13} = 0$) after 250 iterations. From the figure, the channel selection probabilities of the users converge to pure strategy in about 100, 250, and 290 iterations, respectively. In addition, the evolution of number of the users choosing different channels is shown in Fig. 2.4. It is noted that the number of users selecting different channels keeps unchanged in about 250 iterations, which again validates the convergence of the proposed game-theoretic interference mitigation approach.

Fig. 2.3 The evolution of channel selection probabilities for three arbitrarily selected users in Rayleigh fading environment ($N = 5, M = 3, D = 0.005$ and $b = 0.1$)

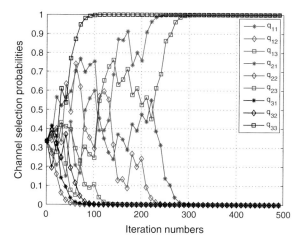

Fig. 2.4 Evolution of the
number of users choosing the
channels in Rayleigh fading
environment
($N = 5, M = 3, D = 0.005$
and $b = 0.1$)

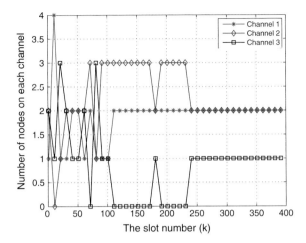

2.5.1.2 Convergence Behavior in Static Environment

In this part, we study the convergence with static channel gains and compare it with an existing static algorithm. There is an efficient distributed channel selection algorithm, called GADIA, which is proposed by Babadi and Tarokh [9] and has been shown to achieve good performance in static systems. According to Proposition 2.7, the learning algorithm in this chapter also converges in static environment. The convergence comparison results of an arbitrary network topology with 20 users and five channels are shown in Fig. 2.5. It is seen that the proposed learning algorithm also converges, as the GADIA algorithm. However, the GADIA algorithm converges rapidly and smoothly. The reasons are: (i) the GADIA algorithm measures the received interference on all channels before a user updates the channel selection strategy, and the

Fig. 2.5 Convergence
behavior comparison in
static environment
($N = 20, M = 5$,
$D = 0.005$ and $b = 0.1$)

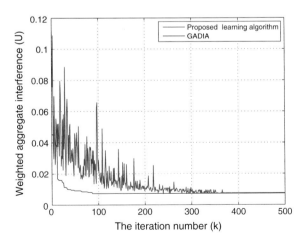

updating procedure is implemented in a deterministic manner, i.e., only one user can update action at a time, whereas (ii) the proposed learning algorithm only measures the received interference on the current chosen channel and the update procedure is implemented in a stochastic manner, i.e., all the users update their actions simultaneously.

2.5.2 Performance Evaluation

2.5.2.1 Performance Comparison for Different Solutions

In this part, the performance of the proposed stochastic automata-based learning algorithm in terms of expected weighted aggregate interference is evaluated. Specifically, we consider a network with five channels, and the number of users increases from 2 to 30. The parameters in the learning algorithm are set as $D = 0.005$ and $b = 0.08$. For comparison, we also consider the following three solutions: the random selection scheme, the worst NE, and the best NE. In the random selection scheme, each user randomly chooses a channel in each slot. Note that the random channel selection seems to be an instinctive method, as the channel gains vary randomly from slot to slot and there is no information exchange. The best (worst) NE are obtained as follows: we run the learning algorithm 10^3 times and then choose the best (worst) result, respectively. According to Theorem 5.1, the best NE is global minimum for the expected weighted aggregate interference.

The comparison results of four solutions is shown in Fig. 2.6. By simulating 10^3 independent trials, the results are obtained by taking the expected value. Some important conclusions can observed: (i) in the under-loaded and equally-loaded scenarios, i.e., $N \leq 5$, the performance of the stochastic learning solution and is almost the same with the best NE, which follows the fact that the global optimum is asymptotically

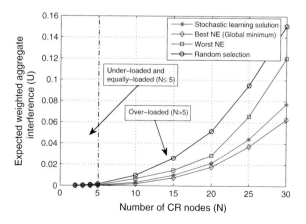

Fig. 2.6 Performance evolution for a distributed network involving in Rayleigh fading environment ($D = 0.005$, $b = 0.08$ and $M = 5$)

achieved, as characterized by Proposition 2.4, and (ii) in the over-loaded network scenarios, i.e., $N > 5$, there is a small performance gap between the learning solution and the best NE. The reason is that the stochastic learning algorithm may converge to an optimal or a suboptimal solution, as characterized by Proposition 2.5, and hence it averagely achieves near-optimal performance. In addition, it is seen that even the worst NE results in less aggregate interference than the random selection scheme. Due to incoordination of the random selection scheme, some channels are crowded whereas others are unoccupied. In comparison, the users choose different channels in pure strategy NE solution, which thus results in lower value of interference.

2.5.2.2 Performance Evaluation for Different Fading Parameters

The performance evaluation for different fading parameters is shown in Fig. 2.7. The presenting results are obtained by simulating 20 topologies with 10^3 independent trials and then taking the average values. No-fading implies that only large-scale power-loss is considered and 0 dB-mean is with unit-mean. From the figure, it can be observed that the performance gap between No-fading and Rayleigh with 0 dB-mean is trivial. According to Proposition 2.6, their performance should be the same as the expected channel gains are the same. Moreover, as the mean value of Rayleigh fading increases, e.g., increasing from 1 to 3 dB, the caused interference increases as can be expected.

2.5.2.3 Performance Evaluation for Different Fading Models

In this part, different fading models are considered. Specifically, the following well-known models including Rayleigh, Nakagami, and Log-normal is considered:

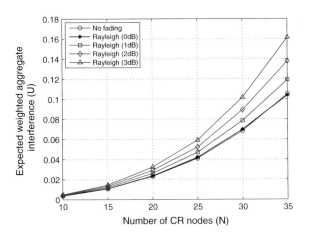

Fig. 2.7 The comparison results of expected aggregate interference for different Rayleigh fading parameters ($D = 0.005$, $b = 0.08$ and $M = 5$)

- In Rayleigh model, the channel gains are exponentially distributed with unit-mean.
- In Nakagami model, the probability distribution function of the channel gains is determined by $f(x) = \frac{m^m x^{m-1}}{\Gamma(m)} e^{-mx}, x \geq 0$.
- In Log-normal model, the channel gains is modeled by a random variable e^X, where X is a Gaussian variable with zero-mean and variance σ^2. Log-normal fading is usually characterized in the dB-spread form which is related to σ, by $\sigma = 0.1 \log(10)\sigma_{dB}$. The dB-spread of Log-normal fading typically ranges from 4 to 12 dB as indicated by the empirical measurements [15].

The comparison results of expected aggregate interference for different fading models are shown in Fig. 2.8. The results are obtained by simulating 20 independent topologies with 10^3 independent trials and then taking the average value. As all the presented fading models are with unit-mean, the interference performance gap is trivial, which directly follows the argument characterized by Proposition 2.6. Also, the comparison results of expected normalized achievable throughput for different fading models are presented in Fig. 2.9. As the number of users increases, the expected normalized achievable rate decreases as expected. Some interesting observations are: (i) Rayleigh fading outperforms Nakagami fading and Log-normal fading, and (ii) the performance of Log-normal fading is almost the same with that of No-fading. We think the reasons may be as follows: (i) multiuser diversity of Rayleigh fading is stronger than those of other fading models, and (ii) the multiuser diversity of Log-normal fading is weak.

Fig. 2.8 The comparison results of expected aggregate interference for different fading models ($D = 0.005$ and $b = 0.08$)

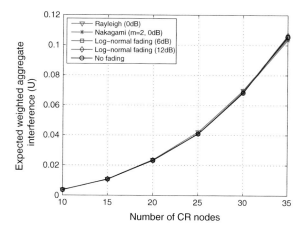

Fig. 2.9 The comparison results of expected normalized achievable throughput for different fading models ($D = 0.005$ and $b = 0.08$)

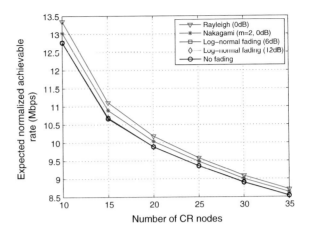

2.6 Concluding Remarks

Compared with previous studies, the key difference in this chapter is that the channel gains are time-varying. In another work [22], we have studied the opportunistic spectrum access problem with time-varying spectrum opportunities, in which the channel states (idle or occupied) change randomly from slot to slot. The stochastic learning automata algorithm was also used therein and its convergence toward pure strategy NE of potential games was rigorously proved. Note that the most promising property of the stochastic learning automata is that the received payoff can be random or deterministic. As a result, we believe that the methodology used in this chapter provides an efficient approach for solving decision-making problems in time-varying environment, which are common in practical wireless networks.

References

1. J. Huang, R. Berry, M. Honig, Distributed interference compensation for wireless networks. IEEE J. Sel. Areas Commun. **24**(5), 1074–1084 (2006)
2. R. Menon, A.B. MacKenzie, J. Hicks et al., A game-theoretic framework for interference avoidance. IEEE Trans. Commun. **57**(4), 1087–1098 (2009)
3. R. Menon, A.B. MacKenzie, R.M. Buehrer et al., Interference avoidance in networks with distributed receivers. IEEE Trans. Commun. **57**(10), 3078–3091 (2009)
4. N. Nie, C. Comaniciu, Adaptive channel allocation spectrum etiquette for cognitive radio networks. Mob. Netw. Appl. **11**(6), 779–797 (2006)
5. Q.D. La, Y.H. Chew, B.H. Soong, An interference-minimization potential game for OFDMA-based distributed spectrum sharing systems. IEEE Trans. Vehic. Technol. **60**(7), 3374–3385 (2011)
6. C. Lăcătuş, C. Popescu, Adaptive interference avoidance for dynamic wireless systems: A game-theoretic approach. IEEE J. Sel. Topics Signal Process **1**(1), 189–202 (2007)

7. Q. Yu, J. Chen, Y. Fan, X. Shen, Y. Sun, Multi-channel assignment in wireless sensor networks: A game theoretic approach, in *Proceedings of the INFOCOM '10*, pp. 1–9 (IEEE, New York, March 2010)
8. Q. Wu, Y. Xu, L. Shen, J. Wang, Investigation on GADIA algorithms for interference avoidance: A game-theoretic perspective. IEEE Commun. Lett. **16**(7), 1041–1043 (2012)
9. B. Babadi, V. Tarokh, GADIA: A greedy asynchronous distributed interference avoidance algorithm. IEEE Trans. Inf. Theory **56**(12), 6228–6252 (2010)
10. J. Wang, Y. Xu, Q. Wu, Z. Gao, Optimal distributed interference avoidance: Potential game and learning. Trans. Emerg. Telecommun. Technol. **23**(4), 317–326 (2012)
11. Q. Wu, Y. Xu, J. Wang, L. Shen, J. Zheng, A. Anpalagan, Distributed channel selection in time-varying radio environment: Interference mitigation game with uncoupled stochastic learning. IEEE Trans. Vehic. Technol. **62**(9), 4524–4538 (2013)
12. L. Cao, H. Zheng, Distributed rule-regulated spectrum sharing. IEEE J. Sel. Areas Commun. **26**(1), 130–145 (2008)
13. L. Garcia, K. Pedersen, P. Mogensen, Autonomous component carrier selection: Interference management in local area environments for LTE-advanced, in *IEEE Communications Magazine*, pp. 110–116 (2009)
14. N. Bambos, Toward power-sensitive network architectures in wireless communications: Concepts, issues, and design aspects. IEEE Personal Commun. **5**(3), 50–59 (1998)
15. G. Stuber, *Princ. Mob. Commun.*, 2nd edn. (Kluwer Academic Publishers, Norwell, 2001)
16. L. Law, J. Huang, M. Liu, S.R. Li, Price of anarchy for cognitive MAC games, in *Proceedings of the IEEE GLOBECOM* (2009)
17. D. Monderer, L.S. Shapley, Potential games. Games Econ. Behav. **14**, 124–143 (1996)
18. J. Marden, G. Arslan, J. Shamma, Joint strategy fictitious play with inertia for potential games. IEEE Trans. Autom. Control **54**(2), 208–220 (2009)
19. Y. Xu, J. Wang, Q. Wu et al., Opportunistic spectrum access in cognitive radio networks: Global optimization using local interaction games. IEEE J. Sel. Topics Signal Process **6**(2), 180–194 (2012)
20. K. Verbeeck, A. Nowé, Colonies of learning automata. IEEE Trans. Syst., Man, Cybern. B **32**(6), 772–780 (2002)
21. P. Sastry, V. Phansalkar, M. Thathachar, Decentralized learning of nash equilibria in multi-person stochastic games with incomplete information. IEEE Trans. Syst., Man, Cybern. B **24**(5), 769–777 (1994)
22. Y. Xu, J. Wang, Q. Wu et al., Opportunistic spectrum access in unknown dynamic environment: A game-theoretic stochastic learning solution. IEEE Trans. Wirel. Commun. **11**(4), 1380–1391 (2012)

Chapter 3
Game-Theoretic MAC-Layer Interference Coordination with Orthogonal Channels

3.1 Introduction

In this chapter, we consider the problem of opportunistic spectrum access in a kind of networks, where the users are spatially located and direct interaction/interference only emerges between neighboring users [1–6]. We investigate this problem from a perspective of interference minimization. Note that the commonly used interference model in the literature is the PHY-layer interference models, in which the focus is to minimize the amount of experienced interference [7]. In methodology, the PHY-layer interference model is more suitable for wireless communication systems with interference channel models, e.g., the code-division multiple access (CDMA) and orthogonal frequency-division multiple access (OFDMA) systems. However, it may not suitable for wireless communication systems with collision channel model, e.g., carrier sensing multiple access (CSMA). In particular, it was recently reported in [8] that the traditional PHY-layer interference model is not applicable for collision channels, e.g., in the 802.11b-based networks.

To capture the mutual interference behavior in multiple access control mechanisms, this chapter considers a new interference metric, called the MAC-layer interference, which is defined as the number of neighboring users choosing the same channel. Compared with the traditional PHY-layer interference model, the MAC-layer interference essentially determines whether two users interfere with each other or not. Based on this definition, we formulate a MAC-layer interference minimization game, and then propose an uncoupled learning algorithm, called the binary log-linear learning algorithm. It is proved that the learning algorithm asymptotically achieves the optimal NE solution and minimizes the aggregate MAC-layer interference. Note that the main analysis and results in this chapter were presented in [9].

© The Author(s) 2016
Y. Xu and A. Anpalagan, *Game-theoretic Interference Coordination*
Approaches for Dynamic Spectrum Access, SpringerBriefs in Electrical
and Computer Engineering, DOI 10.1007/978-981-10-0024-9_3

3.2 Motivation, Definition, and Discussion of MAC-Layer Interference

3.2.1 Motivation and Definition

Due to the open attribute of wireless transmissions, mutual interference is unavoidable in multiuser wireless systems. In the literature, the commonly used model is the PHY-layer interference model [7], in which the focus is to minimize the amount of experienced interference. However, it is noted that the PHY-layer interference model is more suitable for communication systems with interference channel models, e.g., CDMA and OFDMA systems, and not applicable for communication systems with collision channel model, e.g., CSMA and Aloha.

Recently, the experimental results reported in [8] show that for wireless communication systems with CSMA, some interesting features can be observed. In particular, let us consider two nodes (links), which are equipped with 802.11a/b/g cards. It is emphasized here that the considered node (link) actually consists of a transmitter and a receiver located closely [8]. The lognormal fading model is considered, as it addresses the medium-scale path loss well. Specifically, the signal strength (RSS) received at a link from the other link is

$$S = P_t d^{-\beta} e^X, \tag{3.1}$$

where P_t is the transmitting power, d is the physical distance between the two nodes, β is the path loss exponent, and X is a Gaussian variable with zero-mean and variance σ^2. Note that the lognormal fading is usually measured in the dB-spread form which is characterized by $\sigma = 0.1 \log(10) \sigma_{dB}$. As indicated by the empirical measurements [10], the dB-spread of the lognormal fading typically ranges from 4 to 12 dB.

According to the principle of CSMA, a link can hear the transmission of the other link if the received RSS is greater than a threshold S_{th}. In the experiment, we set $P_t = 1$ W, $\beta = 2$, $\sigma_{dB} = 6$ dB, and $S_{th} = 8.1633 \times 10^{-6}$ W. Denote s_1 and s_2 as the achievable throughput of node1 and node2, respectively, when the other node is inactive, and s_1' and s_2' as their achievable throughput when both nodes are active simultaneously. Then, the relationship between the normalized ratio $\gamma = \frac{s_1' + s_2'}{s_1 + s_2}$ and the distance d is used to investigate the effect of interference on the throughput [8]. By simulating 10^6 independent trials and then taking the expected value, we illustratively present the simulation result in Fig. 3.1. Similar to the important observations shown in [8], there are two interesting results:

- The throughput ratio sharply increases from 0.5 (severe interference) to 1 (almost no interference) with a slight increase in the physical distance. As a result, it can be divided into three regions, i.e., interference region, transitional region, and non-interference region, as shown in the figure.

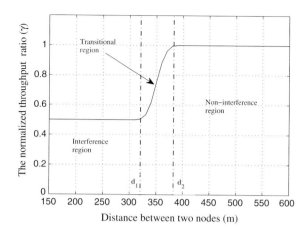

Fig. 3.1 The effect of mutual interference on the normalized throughput ratio

- The throughput ratio in both interference and non-interference regions does not change with the distance, while increasing linearly with the distance in the transitional region.

Some explanations from the perspective of interference are given below: (i) when the two nodes are located in the interference region, only one node can transmit successfully at a time, as they can hear the transmission of each other. In other words, they interfere with each other; (ii) when located in the non-interference region, they can transmit successfully and simultaneously as they do not hear transmission of each other. In other words, there is no interference; and (iii) when located in the transitional region, they can probabilistically hear each other due to the randomness of channel fading. In other words, there exists probabilistic interference, which will be discussed in the subsequent.

As the span of the transitional region $d_2 - d_1$ is relatively small, a simplified interference model can be used to analyze interference among the users. Specifically, if the throughput ratio is less than a threshold, e.g., 0.95, mutual interference between the two nodes exists, and no interference otherwise. Denote the distance corresponding to the interference threshold as d_0, $d_1 < d_0 < d_2$. It motivates us to define the following MAC-layer interference:

$$\alpha = \begin{cases} 1, & x \le d_0 \\ 0, & x > d_0, \end{cases} \tag{3.2}$$

where x is the distance between the two nodes. As a result, the normalized throughput is approximately given by $R = \frac{1}{1+\alpha}$, which provides a good approximation for the measured results [8]. Therefore, although sacrificing a little accuracy, efficient opportunistic spectrum access approaches can be developed using the MAC-layer interference model.

An illustrative comparison of the PHY-layer and MAC-layer interference is presented in Fig. 3.2. In traditional models, the PHY-layer interference is a decreasing

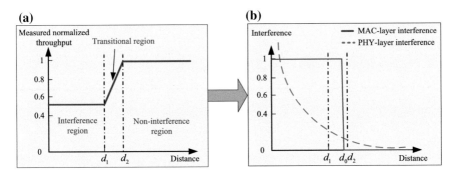

(a)

(b)

Fig. 3.2 The illustrative comparison of the PHY-layer and MAC-layer interference models. **a** The measured normalized throughput versus the distance between two links. **b** The comparison of the proposed MAC-layer interference and traditional PHY-layer interference

function of the physical distance. It is noted from Fig. 3.2b that the PHY-layer interference model does not coincide with the measured results in the context of CSMA. For instance, decreasing the distance of two nodes located in the interference region definitely increases the PHY-layer interference for both; however, it does not lead to decrease in the throughput. Again, the PHY-layer interference models are essentially more suitable for interference channels, while the MAC-layer interference models are more suitable for collision channels.

For multiuser systems, the normalized throughput of user n is given by $\frac{1}{1+\sum_m \alpha_{mn}}$, where α_{mn} is the interference indicator between n and m. Then, the MAC-layer interference experienced by user n in a multiuser system is defined as

$$I_n = \sum_m \alpha_{mn}. \qquad (3.3)$$

3.2.2 Discussion on the Impact of Channel Fading

In this part, we discuss some issues related to channel fading, which is an important attribute of wireless channel. Generally, the received signal strength (RSS) at a node from the node is given by $P_t x^{-\beta} \varepsilon$, where P_t is the transmitting power, x is the physical distance, β is the path loss exponent, and ε is the instantaneous random component of the path loss [10], e.g., lognormal fading and Rayleigh fading. Generally, a link can hear the transmission of the other link if the RSS is greater than a threshold. In the following, we discuss the impact of channel fading on the MAC-layer interference model in the three regions, respectively:

- In the interference region, the large-scale path loss component, i.e., $P_t x^{-\beta}$, is *strong* enough. Thus, one node can deterministically hear the transmission of the other node no matter what are the instantaneous realizations of channel fading. In other words, the impact of channel fading is concealed by the strong large-scale path loss component.

- In the transitional region, the large-scale path loss component is *medium*. Thus, the received RSS is randomly fluctuating around the interference threshold. As a result, one node can probabilistically hear the transmission of the other node.
- In the non-interference region, the large-scale path loss component is *weak*. Thus, one node cannot hear transmission of the other node. In other words, the impact of channel fading is eliminated by the far physical distance.

Remark 3.1 In order to address the interference model in the transitional region more concisely, we can extend the binary interference model to a real-valued one. In particular, an improved MAC-layer interference can be defined as follows:

$$\alpha' = \begin{cases} 1, & x \leq d_1 \\ \frac{x-d_1}{d_2-d1}, & d_1 < x < d_2 \\ 0, & x \geq d_2 \end{cases} \tag{3.4}$$

Note that α' is (3.4) a continuous value ranging in $[0, 1]$. In particular, $\alpha' = 1$ and $\alpha' = 0$ correspond to the same meanings as those in (4.3), while $0 < \alpha' < 1$ corresponds to the probabilistic interference in the transitional region. Similarly, the normalized throughput of a node is then given by $R = \frac{1}{1+\alpha'}$, which fits the measured results well. The real-valued interference model is more precise than the binary interference model characterized by (4.3), as it captures the randomness of channel fading in the transitional region. For presentation of analysis, we only consider the binary interference model in this chapter. Following similar methodology presented in this chapter, the analysis of real-valued MAC-layer interference model can be found in [12].

3.3 System Model and Problem Formulation

3.3.1 Bilateral Interference Networks

Consider a wireless canonical network consisting of N secondary users, in which each user represents a closely located pair of transmitter and receiver [1]. There are M licensed channels owned by the primary users and can be opportunistically used by the secondary users when not occupied. Due to the limited transmitting power of the users, mutual interference only occurs among nearby users [3, 4]. As a result, we can characterize the limited range of interference by an un-directional graph $G = (\mathbb{N}, \mathbb{E})$, where $\mathbb{N} = \{1, \ldots, N\}$ is the vertex set and $\mathbb{E} \subset \mathbb{N} \times \mathbb{N}$ is the edge set. Each vertex represents a user, and the edges correspond to the potential mutual interference relationship among the users when transmitting on the same channel.

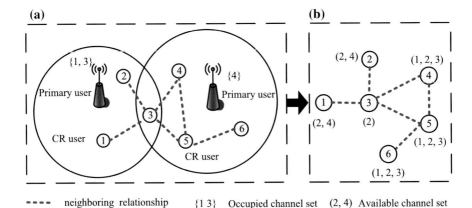

Fig. 3.3 An example of the BI-CRN with four licensed channels. Note that users 1 and 3 interfere with each other when transmitting on the same channel, whereas user 1 will never interfere with user 2 directly. **a** Deployment of the considered BI-CRN **b** The un-directional graph

Although the mutual interference is inherently determined by the received RSS, we can use a distance-relevant interference model as analyzed above. Specifically, if the distance between two users m and n, denoted as D_{mn}, is less than a threshold D_0, they can interfere with each other when simultaneously transmitting on the same channel; thus, m and n are connected by an edge, i.e., there is an edge $e_{mn} = (m, n) \in \mathbb{E}$. For simplicity of analysis, it is assumed that the interference is bilateral between any two users, i.e., user m is also interfered by user n if it interferes with n. We call this kind of networks bilateral interference cognitive radio networks (BI-CRNs).

The spectrum available opportunity is characterized by a channel availability vector C_n, $n \in \mathbb{N}$. In particular, $C_n = \{C_{n1}, C_{n2}, \ldots, C_{nM}\}$, where $C_{nm} = 1$ implies that channel m is available for user n, while $C_{nm} = 0$ means that it is occupied and not available. Note that due to their different localizations, the spectrum opportunities vary from user to user. For simplicity of analysis, it is assumed that spectrum sensing is perfect. Furthermore, the spectrum opportunities are quasi-static in time. Note that such an assumption holds in some realistic networks where the spectrum opportunities are slow-varying, e.g., IEEE 802.22 [11]. An illustrative example of a BI-CRN with six users, two primary users, and four licensed channels is shown in Fig. 3.3.

3.3.2 MAC-Layer Interference Minimization

Due to hardware limitation, it is assumed that all the users can sense all channels simultaneously but transmit on only one channel at a time [13]. Denote $\mathbb{J}_n$ as the neighboring user set of user n, i.e.,

$$\mathbb{J}_n = \{i \in \mathbb{N} : (i, n) \in \mathbb{E}\}, \tag{3.5}$$

Suppose that user n chooses a channel a_n, $a_n \in \{1, \ldots, M\}$, for transmission. Some efficient distributed approaches such as CSMA and distributed TDMA can be applied to coordinate the transmissions among neighboring and interfering users. Thus, the individual achievable throughput of user n under channel selection profile $a = \{a_1, \ldots, a_N\}$ can be expressed by

$$r_n(a_1, \ldots, a_N) = \frac{f(c_n + 1)R_{a_n}}{c_n + 1}, \tag{3.6}$$

where $f(k)$, $0 < f(k) \leq 1$, is the throughput loss function when there are k users competing for a single channel [14], which is decreasing over k. R_{a_n} is the transmission rate of channel a_n, and c_n is the number of neighboring users also choosing the same channel with user n, i.e.,

$$c_n = \sum_{j \in \mathbb{J}_n} \delta(a_n, a_j), \tag{3.7}$$

where $\delta(x, y)$ is the following indicator function:

$$\delta(x, y) = \begin{cases} 1, & x = y \\ 0, & x \neq y. \end{cases} \tag{3.8}$$

Therefore, the network throughput can be expressed as

$$R(a_1, \ldots, a_N) = \sum_{n \in \mathbb{N}} r_n. \tag{3.9}$$

As the decision variables (channel selection) are discrete, the problem of maximizing network throughput is a combinatorial problem on a graph and hence is NP-hard. Motivated by the previous work on minimizing the aggregate PHY-layer interference [7], we consider minimizing the MAC-layer interference in this chapter. Note that the MAC-layer interference experienced by user n is then given by c_n. From the user side, it is desirable to minimize the value of c_n, as minimizing c_n implies maximizing its individual achievable throughput. Thus, from the network side, it is also desirable to minimize the lower aggregate MAC-layer interference. Based on this consideration, we can quantitatively characterize the aggregate MAC-layer interference experienced by all the users as follows:

$$I_g(a_1, \ldots, a_N) = \sum_{n \in \mathbb{N}} c_n \tag{3.10}$$

Consequently, the optimization objective is to find an optimal channel selection a^{opt} such that the aggregate MAC-layer interference is minimized, i.e.,

$$P1: \quad a^{opt} \in \arg\min \; I_g. \tag{3.11}$$

Again, due to the higher-order of computational complexity, $P1$ is hard to resolve even in a centralized manner, not to mention in a distributed manner. Furthermore, the incomplete information constraint, i.e., lack of information about other users, brings about more challenges and difficulties.

3.4 MAC-Layer Interference Minimization Game

Due to the nature of distributed decision-making, we formulate the problem of interference mitigation in BI-CRNs as a non-cooperative game. Different from the game models in the last chapter, the formulated game in this chapter belongs to local interaction games (also known as graphical game) [1], in which the utility of a player only depends on the actions its neighboring users.

3.4.1 Graphical Game Model

Formally, the formulated MAC-layer interference minimization game is denoted as $G = [\mathbb{N}, \{A_n\}_{n\in\mathbb{N}}, \{\mathbb{J}_n\}_{n\in\mathbb{N}}, \{u_n\}_{n\in\mathbb{N}}]$, where $\mathbb{N} = \{1, \ldots, N\}$ is the set of players, $A_n = \{m \in \mathbb{M} : C_{nm} = 1\}$ is the set of player n's available actions (channels), $\mathbb{J}_n$ is the neighboring set of player n, and u_n is its utility function. Generally, in global interactive games, the utility function of each player n is determined by $u(a_n, a_{-n})$, where $a_n \in A_n$ is the chosen action of player n and $a_{-n} \in A_1 \otimes \cdots A_{n-1} \otimes A_{n+1} \ldots A_N$ denotes the action profile all the players except n. However, due to the limited interference in the considered BI-CRNs, the achievable throughput of a player is only determined by its own action as well as the action profile of its neighboring users. Therefore, the utility function can be expressed as $u_n(a_n, a_{\mathbb{J}_n})$, where $a_{\mathbb{J}_n}$ is the action profile of n's neighboring set. In the MAC-layer interference mitigation game, the utility function is defined as follows:

$$u_n(a_n, a_{\mathbb{J}_n}) = L_n - c_n(a_n, a_{\mathbb{J}_n}), \tag{3.12}$$

where $c_n(a_n, a_{\mathbb{J}_n}) \equiv c_n(a_1, \ldots, a_N)$ is the MAC-layer interference experienced by n and L_n is a positive constant satisfying $L_n > |\mathbb{J}_n|$, where $|X|$ denotes the cardinality of set X. Therefore, the purpose of adding L_n in the utility function is to keep the utility function positive, which makes the received payoffs compatible with the proposed learning algorithms. Moreover, L_n can be determined by the users independently and autonomously. As each player in a non-cooperative game maximizes its individual utility function, the proposed game can be expressed as

$$G: \quad \max_{a_n \in A_n} u_n(a_n, a_{\mathbb{J}_n}), \forall n \in \mathbb{N} \tag{3.13}$$

3.4.2 Analysis of Nash Equilibrium

In this part, we analyze the properties of Nash equilibrium (NE) of G in terms of existence of NE and performance bounds.

Theorem 3.1 *The formulated MAC-layer interference mitigation game G is an exact potential game which has at least one pure strategy NE point.*

Proof According to the definition of exact potential game presented in Chap. 1 (See Definition 1.2 therein), we need to prove that there is a potential function such that the change in the utility function caused by the unilateral action deviation of an arbitrary player is the same as that in the potential function. For the formulated MAC-layer interference mitigation game, the following potential function is constructed:

$$\Phi(a_n, a_{-n}) = -\frac{1}{2} \sum_{n \in \mathbb{N}} c_n(a_1, \dots, a_N). \tag{3.14}$$

Now, suppose that there is an arbitrary player n unilaterally changing its channel selection from a_n to $\bar{a}_n$. Following the similar lines given in our previous work [1, 2], it can be verified that the following equation always holds:

$$\Phi(\bar{a}_n, a_{-n}) - \Phi(a_n, a_{-n}) = u_n(\bar{a}_n, a_{\mathbb{J}_n}) - u_n(a_n, a_{\mathbb{J}_n}), \tag{3.15}$$

which shows that the MAC-layer interference mitigation game G is an exact potential game. Thus, Theorem 5.1 follows. □

As the users in the non-cooperative games are selfish, the NE solutions maybe inefficient, which is known as *tragedy of commons* [15]. In the following, we investigate the performance bounds of NE solutions of the MAC-layer interference game. To begin with, the aggregate MAC-layer interference of a pure strategy NE $a_{\mathrm{NE}} = \{a_1^*, \dots, a_N^*\}$ is given by

$$U(a_{\mathrm{NE}}) = \sum_{n \in \mathbb{N}} c_n(a_n^*, a_{\mathbb{J}_n}^*) \tag{3.16}$$

Generally, the MAC-layer interference mitigation game G may have multiple pure strategy NE points but the number is hard to calculate [16]. The following theorems characterize the performance bounds of the game.

Theorem 3.2 *The aggregate MAC-layer interference of any pure strategy NE solution is bounded by $U(a_{NE}) \leq \sum_{n=1}^{N} \frac{|\mathbb{J}_n|}{|A_n|}$, for any network topology and spectrum opportunities.*

Proof Refer to [9]. □

From Theorem 4.2, it is known that the larger number of available channels ($|A_n|$) and smaller number of neighboring users ($|\mathbb{J}_n|$) are preferable, as can be expected in any multiuser multichannel networks. In particular, the performance bound can be refined for some special kinds of systems.

Proposition 3.1 *If all the channels are available to each user, then the aggregate MAC-layer interference at any NE solution is bounded by $U(a_{NE}) \leq \frac{2N}{M}$.*

Proof Since all the channels are available to each user, we have $|A_n| = M, \forall n \in \mathbb{N}$. Thus, the following equation follows:

$$U_{(a_{NE})} \leq \sum_{n=1}^{N} \frac{|\mathbb{J}_n|}{M}, \tag{3.17}$$

which can be straightforwardly obtained by Theorem 4.2 directly. Moreover, it can be verified that the following always holds for any network topology:

$$\sum_{n=1}^{N} |\mathbb{J}_n| = 2N \tag{3.18}$$

Now, combining (3.18) and (3.17) proves this proposition. □

Theorem 3.3 *The best pure strategy NE point of G is a global minimum of the MAC-layer interference mitigation problem P1.*

Proof According to (3.14), the potential function of the formulated game and the aggregate MAC-layer interference are related by $\Phi(a_n, a_{-n}) = -\frac{1}{2} I_g(a_n, a_{-n})$. Thus, we have

$$a^{opt} \in \arg\max \ \Phi(a_n, a_{-n}). \tag{3.19}$$

which is obtained from (3.11). That is, any channel selection profile minimizing the aggregate MAC-layer interference maximizes the potential function. Recalling the important property of potential game, i.e., any global or local maximizer of the potential function constitutes a pure strategy NE point [17], it is known that the best pure strategy NE point is a global minimum of $P1$, which proves Theorem 3.3. □

The result shown in Theorem 3.3 is interesting and promising, since the global optimality emerges as the result of distributed and selfish decisions via game design and optimization.

Algorithm 2*: the binary log-linear learning algorithm*

Initialization: Let each player i randomly select a channel from its available channel set, i.e., $a_i(0) \in A_i, \forall i \in \mathbb{N}$.

Loop for $k = 0, 1, 2, \ldots,$

1. Player selection: Using the 802.11 DCF-like coordination mechanism, a player, say n, is randomly selected in an autonomous manner. Then, all the users adhere to their channel selections in an estimation period and the chosen user estimates its received utility $u_n(k)$.

2. Exploration: Player n randomly chooses a channel $m \in A_n, m \neq a_n(k)$. Then, all the users adhere to their selections in the subsequent estimation period and the chosen player estimates its received utility in channel m, which is denoted as v_m.

3. Updating channel selection: The chosen player n updates its channel selection strategy using the following log-linear rule:

$$\Pr[a_n(k+1) = m] = \frac{\exp\{v_m \beta\}}{\exp\{v_m \beta\} + \exp\{\hat{u}_n(k)\beta\}}$$

$$\Pr[a_n(k+1) = a_n(k)] = \frac{\exp\{u_n(k)\beta\}}{\exp\{v_m \beta\} + \exp\{\hat{u}_n(k)\beta\}}, \tag{3.20}$$

where β is a learning parameter. Meanwhile, all other players keep their selections unchanged, i.e., $a_{-n}(k+1) = a_{-n}(k)$.

End loop

3.5   The Binary Log-Linear Learning Algorithms for Achieving Best NE

3.5.1   Algorithm Description

As the MAC-layer interference mitigation problem is now formulated as an exact potential game, there are large number of learning algorithms to achieve pure strategy NE, e.g., best response dynamic [17], spatial adaptive play [1], and fictitious play [18]. However, these algorithms belong to coupled algorithms and hence need to know information about other players. Although the stochastic automata learning algorithm, which was applied in the last chapter, is uncoupled, it may converge to a suboptimal solution. Thus, in this chapter, an uncoupled and optimal distributed learning algorithm, called the binary log-linear learning algorithm, is applied to achieve the best NE.

The binary log-linear learning algorithm is described in Algorithm 2. The key idea is that only a player is randomly chosen to explores the channels. Based on the explanation results, the player updates its selection using the log-linear rule. Some practical concerns of Algorithm 1 are discussed as follows: (i) in the step of player selection, the selection of an autonomous and random player can be achieved using a 802.11 DCF-like contention mechanisms over a common control channel (CCC)

[1, 19], and (ii) the following stop criterions can be used: (i) the maximum iteration number is reached, and (ii) the variation of the achieved utility during a certain period is trivial.

Note that the utility function, i.e., the experienced MAC-layer interference, cannot be measured directly by the users; we use the following simple method to estimate the MAC-layer interference experienced by a user. Of course, other more practical but also completed methods for estimating the number of competing users can also be used, e.g., [23, 24]. The reason we use such a simple method is our focus in designing game-theoretic distributed MAC-layer interference mitigation approach but not the estimation algorithms. Specifically, suppose that there are total H slots in each estimation period and T_n is the number of slots in which user n successfully access the channel. As a result, the maximum-likelihood estimation (MLE) of the MAC-layer interference experienced by user n can be calculated by

$$\hat{s}_n = \frac{H}{T_n} - 1,$$ (3.21)

which further implies that the MLE of the received payoff in an estimation period is expressed as

$$\hat{u}_n = L_n + 1 - \frac{H}{T_n}$$ (3.22)

Based on the above estimation approach, an illustrative diagram of the binary log-linear learning algorithm for the formulated MAC-layer interference mitigation game is shown in Fig. 3.4.

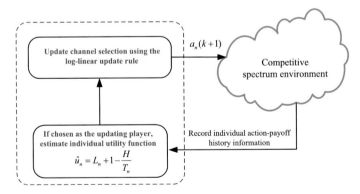

Fig. 3.4 The illustrative diagram of the binary log-linear learning algorithm for the formulated MAC-layer interference mitigation game

3.5.2 *Convergence Analysis*

Let $\mathbb{A}$ be the set of available channel profiles of all the players, i.e., $\mathbb{A} = A_1 \otimes \cdots \otimes A_N$, then the properties of the binary log-linear learning algorithm are characterized by the following theorems.

Theorem 3.4 *For the binary log-linear learning algorithm, the unique stationary distribution $\mu(a) \in \Delta(\mathbb{A})$ of any channel selection profile $s \in \mathbb{A}$, $\forall \beta > 0$, is given as*

$$\mu(a) = \frac{\exp\{\beta \Phi(a)\}}{\sum_{s \in \mathbb{A}} \exp\{\beta \Phi(s)\}}, \tag{3.23}$$

where $\Phi(\cdot)$ is the potential function given in (3.14).

Proof The proof follows the methodology presented in [1, 20–22]. Detailed lines are not presented here and can be found in [9]. $\square$

Theorem 3.5 *With a sufficiently large β, the binary log-linear learning algorithm asymptotically minimizes the aggregate MAC-layer interference I_g.*

Proof Based on Theorem 3.4 and the similar lines for proof presented in our previous work [1] (see Theorem 4 therein), it can be proved that when β goes sufficiently large, the binary log-linear algorithm asymptotically converges to a channel selection profile that maximizes the potential function. Now, applying again the relationship between the potential function and the original optimization objective, i.e., $\Phi(a) = -\frac{1}{2} I_g(a)$, Theorem 3.5 can be obtained. $\square$

3.6 Simulation Results and Discussion

In this section, simulation results are presented to validate the proposed game-theoretic distributed channel selection solution for MAC-layer interference mitigation. Although the game formation and learning algorithm are only theoretically analyzed for scenarios with no fading, it is shown by simulation results that the proposed game-theoretic solution is also suitable for scenarios with fading. Also, it is suitable for scenarios with both bilateral and unilateral interferences.

3.6.1 *Scenario Setup*

In the simulation study, the users are randomly located in a region. For simplicity, it is assumed that the idle probabilities of all the channels are the same, which is denoted as θ, $0 < \theta < 1$, and the spectrum opportunities are randomly generated according to the idle probabilities independently. However, it should be pointed out that the spectrum opportunities are quasi-static, i.e., they vary slowly in time, or are static during the convergence of the learning algorithm. Furthermore, we assume that different channels support the same transmission rate $R = 1\text{Mbps}$ for the users. The users use a perfect CSMA/CA mechanism to share the idle channels. As a result, the achievable throughput of a user is approximately determined by $\frac{R}{c_n+1}$, which is obtained by setting the throughput loss function to be one, i.e., $f(c_n + 1) \approx 1$ in (4.7).

3.6.2 Scenario with No Fading

In this subsection, we consider scenarios with no fading, where only the large-scale path loss is considered. The learning parameter in the payoff-based log-linear learning algorithm is set to $\beta = 10 + k/50$, where k is the iteration number. In all simulations, the estimation period is set to $H = 100$.

3.6.2.1 Convergence Behavior

In this part, the convergence behavior of the proposed learning algorithm is studied. Specifically, we study a small network as shown in Fig. 3.5, which involves nine CR users and three channels. A scenario with all channels being available for the users is considered. In such a scenario, it provides with the same spectrum opportunities and hence the expected convergence behavior of the learning algorithms can be studied by taking independent trials and then taking the expected results. For the presented network, the expected convergence behaviors of the proposed learning algorithm are shown in Fig. 3.6, which are obtained by simulating 1000 independent trials and then taking the average results. It is noted that the proposed learning algorithm converges

Fig. 3.5 The simulated small network. (Each *circle* represents a CR user, and the *dashed lines* represent the bilateral interferences between the users)

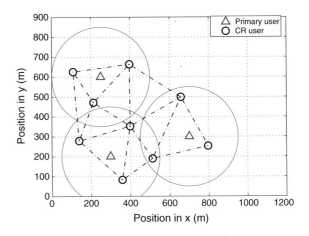

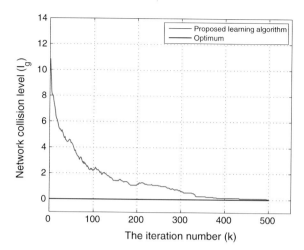

Fig. 3.6 The expected convergence behaviors of the proposed learning algorithms

to the global optimum in about 400 iterations. The results validate the asymptotical optimality of the proposed learning algorithm.

In the following, the effect of the estimate interval H on the convergence of the proposed learning algorithm is shown in Fig. 3.7. The results show that there is a tradeoff between speed and performance with regard to the estimate interval H. It is noted from the figure that larger H leads to higher estimation accuracy while leading to relatively slower convergence speed, as can be expected. Thus, the choice of the estimate period H is application-dependent in practice.

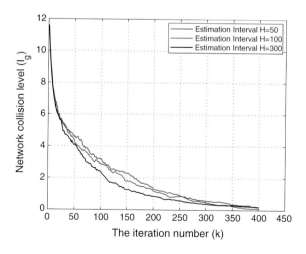

Fig. 3.7 The effect of the estimate interval H on the convergence of proposed learning algorithm ($\beta = 10 + k/50$)

3.6.2.2 Throughput Performance

In this part, we compare the achievable throughput of five methods: throughput maximization using exhaustive search (TMax-ES), random selection, interference minimization using the proposed learning algorithm (Imin-Proposed), interference minimization using spatial adaptive play (Imin-SAP) [1], and interference minimization using best response (Imin-BR) [17]. In the TMax-ES approach, the aggregate network throughput as characterized by (3.9) is directly maximized using an exhaustive search method. In the random selection scheme, each user randomly chooses a channel from its available channel set. In comparison, SAP and BR are two commonly coupled learning algorithms for potential games, which need the information of other users. In addition, the SAP algorithm asymptotically converges to the global maxima of the potential function, while the BR algorithm achieves its global or local maxima randomly.

   (i) Small networks For the small network (see Fig. 3.5), the comparison results of the expected network throughput are shown in Fig. 3.8. It is noted from the figure that the proposed learning algorithm achieves higher network throughput than the BR algorithm and the random selection approach. Furthermore, the throughput gap increases as the channel idle probability θ increases. The reason is that in NE solutions of the game, the users are spread over different channels and hence there is less interference (collision) among the users. Again, this result is due to the fact that all pure strategy NE points of the game minimize the aggregate MAC-layer interference globally or locally, as characterized by Theorems 5.1, 4.2, and 3.3.

   It is noted that the proposed learning algorithm achieves the same performance with that of Imin-SAP. The SAP algorithm is an efficient learning algorithm for potential game, as it asymptotically maximizes the potential function [1], i.e., minimizing the aggregate MAC-layer interference in the formulated MAC-layer interference mitigation game. In comparison, the proposed learning algorithm does not need

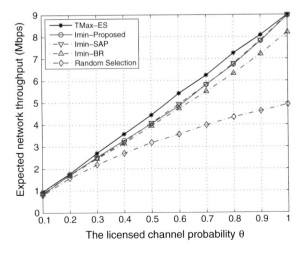

Fig. 3.8 Companion results of five channel selection methods for the simulated small network with no fading (The learning parameter of the learning algorithms is set to $\beta = 10 + k/50$)

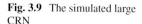

Fig. 3.9 The simulated large CRN

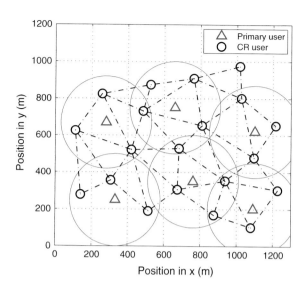

any information about other players while the SAP needs information about other players. In addition, although the proposed learning asymptotically minimizes the aggregate MAC-layer interference, there is a throughput gap between Imin-Proposed and TMax-ES. The reason is as follows: although a lower aggregate MAC-layer interference would lead to higher network throughput as can be expected, a quantitative characterization between minimizing the MAC-layer interference and maximizing the network throughput directly is hard to obtain. Even so, the proposed learning algorithm is desirable for practical applications, as it achieves higher network throughput.

(ii) Large networks We consider a relatively large network as shown in Fig. 3.9, which consists of 20 users and three channels. Figure 3.10 shows the comparison results of the expected network throughput of different solutions. Due to intolerable complexity, the TMax-ES method cannot be applied in this scenario. It is noted that the throughput performance of the proposed learning algorithm is very close to the Imin-SAP approach. These results validate that the proposed learning algorithm is also for large networks.

3.6.3 Scenario with Fading

As fading is common in wireless networks, we study the performance of the proposed learning algorithm in scenarios with lognormal fading. The parameters are set as follows: the transmitting power of all users is $P_t = 1$ W, the path loss exponent is $\beta = 2$, and the dB-spread is 6 dB. Moreover, it is assumed that the detection threshold of CSMA is 8.1633×10^{-6} W. Simulation results show that the proposed learning algorithm also converge for channels with lognormal fading. However, since their

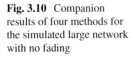

Fig. 3.10 Companion results of four methods for the simulated large network with no fading

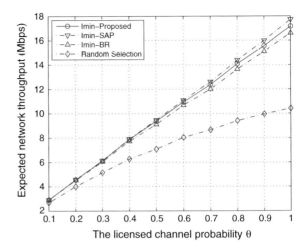

convergence trend is similar to that presented in Fig. 3.6, it is not presented here in order to avoid unnecessary repetition.

For the considered small and large networks, the comparison results for the achievable network throughput are shown in Figs. 3.11 and 3.12, respectively. It is noted that the proposed algorithm also achieves the same performance with that of Imin-SAP, and outperforms both the BR approach and random selection approach. These results validate the effectiveness of the proposed learning algorithm for scenarios with fading.

Fig. 3.11 Throughput companion for the simulated small network with lognormal fading ($P_t = 1$ W, $\beta = 2$ and $\sigma_{dB} = 6$ dB)

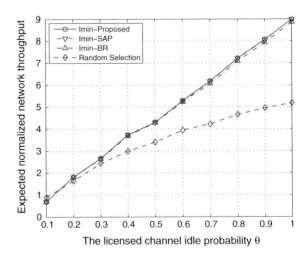

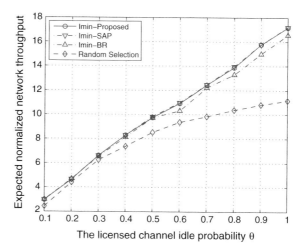

Fig. 3.12 Throughput companion for the simulated large network with lognormal fading ($P_t = 1$ W, $\beta = 2$ and $\sigma_{dB} = 6$ dB)

3.7 Extension to Unilateral Interference CRNs

The above presented analysis is for bilateral interference networks. However, we would like to point out that there are some scenarios involving unilateral interference relationships among the users, e.g., cognitive ad hoc networks. For those networks, the mutual interference relationships can be characterized by a directional graph rather than an undirected graph. We call this kind of networks, unilateral interference cognitive radio networks (UI-CRNs).

3.7.1 System Model

To make it more practical, we then extend BI-CRNs to UI-CRNs in this section. For the unilateral interference networks, a CR receiver suffers from interference from other CR transmitters if the distance between them is less than a predefined threshold, D_I. For simplicity of analysis, we assume that the available spectrum opportunities are identical for any CR transmitter and its dedicated receiver. Then, the heterogeneous spectrum opportunities are also characterized by the channel availability vectors C_n, as discussed before.

The interference relationship is now characterized by a directional graph $\mathbf{G}_d = (\mathbb{N}, \mathbb{E})$. The graph $\mathbf{G}_d$ consists of a set of nodes which is exactly the CR user set $\mathscr{N}$, and a set of edges $\mathscr{E} \subset \mathbb{N}^2$. Denote each edge as an ordered pair (i, j); then, if there is an edge from user i to j, i.e., $(i, j) \in \mathscr{E}$, it means that the transmission of node i interferes with node j when simultaneously transmitting on the same channel. An example of the deployment for the considered unilateral interference cognitive

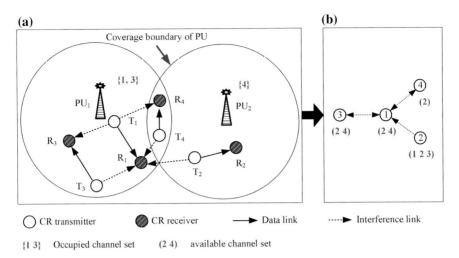

(a) **(b)**

CR transmitter   CR receiver   Data link   Interference link

{1 3}   Occupied channel set   (2 4)   available channel set

Fig. 3.13 An example of the considered UI-CRN with four CR links. (In such a network, the interference is unilateral, e.g., CR link 1 does not interfere with link 2 but it is interfered by link 2.) **a** Deployment of the considered UI-CRN **b** The interference graph

radio network is shown in Fig. 3.13a and the corresponding directional interference graph is shown in Fig. 3.13b.

Following the same methodology used for BI-CRNs, similar definitions for UI-CRNs can also be given. Then, a similar network collision minimization game can be established accordingly. For reducing unnecessary repetition, they are not presented here. For UI-CRNs, the network-centric goal is also to minimize the aggregate MAC-layer interference. However, due to the unilateral interference relationship, it is no longer a potential game. This makes the analysis of the convergence of the proposed uncoupled learning algorithms a formidable task and an open problem.

3.7.2 Simulation Results

Although there is a lack of theoretic analysis, as discussed above, we evaluate the performance of the proposed uncoupled algorithms by simulation study. Specifically, the deployment of the considered UI-CRN is shown in Fig. 3.14. For the considered UI-CRN, the expected convergence behaviors of the proposed learning algorithm are shown in Fig. 3.15. The results are obtained by simulating 1000 independent trials and then taking the average. It is noted that it asymptotically converges to global optimum in about 450 iterations.

Fig. 3.14 An unilateral interference CRN with nine CR users (links) and three licensed channels (*Each circle* represents a CR link, *double arrows* represent bilateral interferences, and *single arrows* represent unilateral interferences)

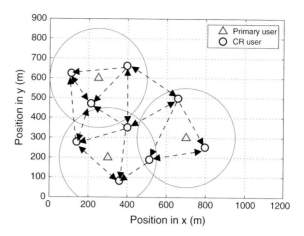

The comparison results of expected network throughput obtained using five methods (exhaustive search, random selection, the proposed algorithm, SAP, and BR) are shown in Fig. 3.16. As for the UI-CRNs, it is also noted from the figure that the proposed learning algorithm achieves higher network throughput than the BR algorithm and the random selection approach. Also, it achieves almost the same throughput with the SAP algorithm. Therefore, we claim that the proposed learning algorithms are not only suitable for BI-CRNs, but also suitable for UI-CRNs, although it lacks rigorous theoretical analysis.

Fig. 3.15 The convergence behaviors of the two algorithms for the simulated UI-CRN ($H = 200$)

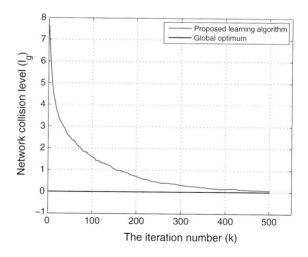

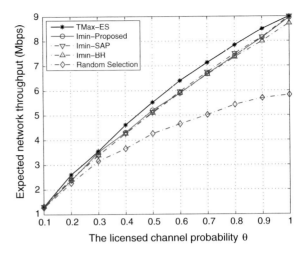

Fig. 3.16 Companion results of four channel selection methods (For the learning algorithms, we set $H = 200$)

3.8 Concluding Remarks

Compared with traditional PHY-layer interference model, e.g., the one considered in Chap. 2, the MAC-layer interference model in this chapter coincides with the experiment results for collision channel models. Moreover, it admits mathematical tractability as it only cares about whether two users interfere with each other or not. Also, the binary log-linear learning algorithm can achieve the best NE of exact potential games. Thus, the results presented in this chapter provide efficient distributed solutions for resource allocation problems over graph/network.

References

1. Y. Xu, J. Wang, Q. Wu et al., Opportunistic spectrum access in cognitive radio networks: Global optimization using local interaction games. IEEE J. Sel. Signal Process **6**(2), 180–194 (2012)
2. Y. Xu, Q. Wu, J. Wang et al., Distributed channel selection in CRAHNs with heterogeneous spectrum opportunities: a local congestion game approach. IEICE Trans. Commun. **E95–B**(3), 991–994 (2012)
3. H. Li, Z. Han, Competitive spectrum access in cognitive radio networks: Graphical game and learning, in *Proceedings of the IEEE WCNC 2010* (2010)
4. M. Azarafrooz, R. Chandramouli, Distributed learning in secondary spectrum sharing graphical game, in *Proceedings of the IEEE GLOBECOM*, pp. 1–6 (2011)
5. M. Liu, et al, Congestion games with resource reuse and applications in spectrum sharing, *GameNets*, pp. 171–179 (2009)
6. C. Peng, H. Zheng, B. Zhao, Utilization and fairness in spectrum assignemnt for opportunistic spectrum access. Mob. Netw. App. **11**(4), 555–576 (2006)
7. C. Lacatus, D. Popescu, Adaptive interference avoidance for dynamic wireless systems: A game-theoretic approach. IEEE J. Sel. Topics Signal Process **1**(1), 189–202 (2007)
8. Y. Ding, Y. Huang, G. Zeng, L. Xiao, Using partially overlapping channels to improve throughput in wireless mesh networks. IEEE Trans. Mob. Comput. **11**(11), 1720–1733 (2012)

9. Y. Xu, Q. Wu, L. Shen, J. Wang, A. Anpalgan, "Opportunistic spectrum access with spatial reuse: Graphical game and uncoupled learning solutions". IEEE Trans. Wirel. Commun. 12(10), 4814–4826 (2013)

10. G. Stuber, *Principles of Mobile Communications*, 2nd edn. (Kluwer Academic Publishers, Norwell, 2001)

11. D. Niyato, E. Hossain, Z. Han, Dynamic spectrum access in IEEE 802.22-based cognitive wireless networks: A game theoretic model for competitive spectrum bidding and pricing. IEEE Wirel. Commu. 16(2), 16–23 (2009)

12. Q. Zhao, L. Shen, C. Ding, Stochastic MAC-layer interference model for opportunistic spectrum access: A weighted graphical game approach, IEEE/KICS J. Commun. Netw. (to appear)

13. J. Jia, Q. Zhang, X. Shen, HC-MAC: A hardware-constrained cognitive MAC for efficient spectrum management. IEEE J. Sel. Areas Commun. 26(1), 466–479 (2008)

14. Y. Xu, J. Wang, Q. Wu et al., Opportunistic spectrum access in unknown dynamic environment: A game-theoretic stochastic learning solution. IEEE Trans. Wirel. Commun. 11(4), 1380–1391 (2012)

15. H. Kameda, E. Altman, Inefficient noncooperation in networking games of common-pool resources. IEEE J. Sel. Areas Commun. 26(7), 1260–1268 (2008)

16. R. Myerson, *Game Theory: Analysis of Conflict* (Harvard University Press, Cambridge, 1991)

17. D. Monderer, L.S. Shapley, Potential games. Games Econ. Behav. 14, 124–143 (1996)

18. J. Marden, G. Arslan, J. Shamma, Joint strategy fictitious play with inertia for potential games. IEEE Trans. Autom. Control 54(2), 208–220 (2009)

19. N. Nie, C. Comaniciu, Adaptive channel allocation spectrum etiquette for cognitive radio networks. Mob. Netw. Appl. 11(6), 779–797 (2006)

20. H.P. Young, *Individual Strategy and Social Structure* (Princeton University Press, New Jersey, 1998)

21. J. Marden, G. Arslan, J. Shamma, Cooperative control and potential games. IEEE Trans. Syst., Man, Cybern. B, 39(6), pp. 1393–1407 (2009)

22. J. Wang, Y. Xu, A. Anpalagan et al., Optimal distributed interference avoidance: Potential game and learning. Trans. Emerg. Telecommun. Technol. 23(4), 317–326 (2012)

23. Y. Xu, Zhan Gao, Qihui Wu, et al, Collision mitigation for cognitive radio networks using local congestion game, in *Proceedings of International Conference on Communication Technology and Application* (2011)

24. A. Toledo, T. Vercauteren, X. Wang, Adaptive optimization of IEEE 802.11 DCF based on bayesian estimation of the number of competing terminals. IEEE Trans. Mob. Comput. 5(9), 1283–1296 (2006)

Chapter 4
Game-Theoretic MAC-Layer Interference Coordination with Partially Overlapping Channels

4.1 Introduction

Due to the nonideal shaping of filter in practical wireless networks, the transmitted signal in a channel always leads to spectrum leak in the adjacent channels. Thus, in order to avoid adjacent channel interference, there should be enough separation between two operational channels. These channels are called orthogonal channels (non-overlapping channels), as adjacent channel interference can be completed eliminated. Most existing dynamic spectrum access approaches focused on assigning non-overlapping channels to the users, e.g., [1–5]. As a result, the basic principle is that only when allocated with non-overlapping channels, two interfering players can transmit simultaneously. However, due to the restriction that two operational channels need to be orthogonal, the number of available channels is very limited. For example, there are total eleven channels but only three orthogonal channels in the IEEE 802.11b networks. As the network becomes dense, mutual interference among the users cannot be completely eliminated, resulting in severe co-channel interference and throughput drop. Therefore, it is important to break the restriction of orthogonal channels and develop new channel usage mechanisms to improve the network throughput.

In this chapter, we consider the problem of dynamic spectrum access with partially overlapping channels (POC), in which the operational channels may be orthogonal or not [6]. As can be expected, the number of operational channels increases significantly when the new paradigm of partially overlapping channels (orthogonal channels) is employed. In partially, it has been shown that the full-range spectrum utilization can be improved significantly using POC [7–14]. The idea of using POC is interesting and promising. However, in most existing researches, the channels are always allocated using centralized approaches, e.g., graph coloring [7], genetic algorithm [13], and other optimization technologies [10, 14]. As stated before, the centralized approach requires a central controller and information exchange among users, which is not feasible in some scenarios. In this chapter, based on the

© The Author(s) 2016
Y. Xu and A. Anpalagan, *Game-theoretic Interference Coordination
Approaches for Dynamic Spectrum Access*, SpringerBriefs in Electrical
and Computer Engineering, DOI 10.1007/978-981-10-0024-9_4

MAC-layer interference model presented in Chap. 3, we formulate an interference game for dynamic spectrum access with POC and propose a distributed learning algorithm to achieve the global optimum. Note that the main analysis and results in this chapter were presented in [15].

4.2 Interference Models and Problem Formulation

Similar to the system model in Chap. 3, we also consider a wireless canonical network which consists of K users. In partially, we take the IEEE 802.11b-based network as the research instance, which were always utilized as the hardware platforms for dynamic spectrum access [3]. In the 802.11b-based networks, the heading access point (AP) and the associated clients form a basic service set (BSS). The AP chooses the operation channel and then all the clients belonging to the same BSS operate on the chosen channel, using the well-known distributed coordination (DCF) mechanism. Denote the available channel set as $\{c_1, \ldots, c_M\}$. Then, the optimization objective for each AP is to choose appropriate operational channel to eliminate mutual interference with other APs.

4.2.1 MAC-Layer Interference Model with Partially Overlapping Channels

For 802.11b-based wireless networks, there are total eleven channels and each one is with 5 MHz bandwidth. It has been reported by measurement results that spectrum leak in IEEE 802.11-based networks is quite severe [13, 14]. Specifically, the power mask on a channel with center frequency F_c can be approximately expressed as [13]

$$P(f) = \begin{cases} 0\,\text{dB}, & |f - F_c| \leq 11\,\text{MHz} \\ -30\,\text{dB}, & 11\,\text{MHz} < |f - F_c| < 22\,\text{MHz} \\ -50\,\text{dB}, & |f - F_c| > 22\,\text{MHz} \end{cases} \tag{4.1}$$

An illustrative diagram of the power mask is shown in Fig. 4.1. Thus, in order to make the mutual interference between two channels to be ignored (less than -30 dB), the separation of the center frequencies should be larger than 22 MHz. That is, the channel number separation of any two orthogonal channels is at least 5. As a result, there are at most three non-overlapping channels available in the IEEE 802.11-based networks. Traditionally, the problem of interference mitigation with orthogonal channel assignment has been well studied in the literature. However, it is also noted that the spectrum resources are not fully exploited since only a small part of channels (orthogonal channels) is utilized. To improve the spectrum efficiency, using partially overlapping channels has been regarded as an interesting and promising approach.

Fig. 4.1 The power mask in 802.11b-based networks

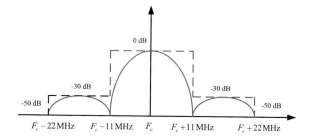

To begin with, we analyze the interference model for partially overlapping channels. It was shown in [13] that the interference in IEEE 802.11-based WLAN is jointly determined by the physical distance and the channel separation. In particular, a binary interference property can be observed with respect to their physical distance [13]. Thus, the MAC-layer interference model presented in Chap. 3 is used in this chapter. If two users choose channels a_1 and a_2, then the channel separation in terms of channel number is calculated as

$$\delta = |a_1 - a_2|. \tag{4.2}$$

For example, $\delta = 2$ if channels 1 and 3 are chosen. Following the similar lines for analysis Chap. 3, an interference range $R_I(\delta)$ can be defined for a specific channel separation distance δ. Specifically, if two users are located in the interference range of each other, they interfere with each other and hence share the channel equally; also, when located beyond the interference range, there is no mutual interference so they can transmit simultaneously.

Without loss of generality, the communication range of a user is denoted as R. Based on the measurement reported in [13], the relationship between the interference range $R_I(\delta)$ and the channel separation δ is presented in Table 4.1. Take the data rate of 2 Mb/s as an illustrative example: (i) $R_I(0) = 2R$, if $\delta = 0$. That is, if the same channel is used by two users, the interference range is then two times the communication range. (ii) $R_I(5) = 0$, if $\delta = 5$. That is, if two orthogonal channels are used, the interference range is zero. (iii) when the channel separation δ increases from 0 to 5, the interference range $R_I(\delta)$ decreases monotonically from $2R$ to 0. For other data rates, e.g., 5.5 and 11 Mb/s, the same trend can also be observed.

Table 4.1 The interference range for different channel separations

Channel separation (δ)	0	1	2	3	4	5
Interference range (2 Mb/s)	2R	1.125R	0.75R	0.375R	0.125R	0
Interference range (5.5 Mb/s)	2R	R	0.625R	0.375R	0.125R	0
Interference range (11 Mb/s)	2R	R	0.5R	0.345R	0.125R	0

Based on the above analysis, the MAC-layer interference between two users is defined as follows:

$$\alpha = \begin{cases} 1, & d \leq R_I(\delta) \\ 0, & d > R_I(\delta) \end{cases} \tag{4.3}$$

where d is the distance between the two users. Thus, the achieved normalized throughput of a user can be expressed as $T = \frac{1}{1+\alpha}$. Furthermore, it can be extended to the scenarios of multiple users. Specially, denote the physical distance between users k and n as d_{kn}, and their channel separation as δ_{kn}. Then, the interference between user k and n is given by

$$\alpha_{kn} = \begin{cases} 1, & d_{kn} \leq R_I(\delta_{kn}) \\ 0, & d_{kn} > R_I(\delta_{kn}) \end{cases} \tag{4.4}$$

Based on the above definition, the achieved normalized throughput of user k is given by $T_k = \frac{1}{1+I_k}$, where $I_k = \sum_{n \neq k} \alpha_{nk}$ can be regarded as the aggregate MAC-layer interference experienced by user k. Similar to Chap. 3, we consider minimizing the aggregate MAC-layer interference with partially overlapping channels in this chapter.

4.2.2 Problem Formulation

Denote the user set as $\mathbb{K}$, i.e., $\mathbb{K} = \{1, 2, \ldots, K\}$. For an arbitrary user k, denote the neighboring user set in the interference range of channel separation $\delta = 4$ as $\mathbb{J}_k^{(4)}$, i.e., $\mathbb{J}_k^{(4)} = \{j \in \mathbb{K} : d_{jk} \leq R_I(4)\}$, where $R_I(4)$ is given in Table 4.1. Furthermore, define the neighboring user set in the range of $R_I(3)$ but beyond the range of $R_I(4)$ as $\mathbb{J}_k^{(3)} = \{j \in \mathbb{K} : R_I(4) < d_{jk} \leq R_I(3)\}$. Similarly, $\mathbb{J}_k^{(2)} = \{j \in \mathbb{K} : R_I(3) < d_{jk} \leq R_I(2)\}$, $\mathbb{J}_k^{(1)} = \{j \in \mathbb{K} : R_I(2) < d_{jk} \leq R_I(1)\}$ and $\mathbb{J}_k^{(0)} = \{j \in \mathbb{K} : R_I(1) < d_{jk} \leq R_I(0)\}$. Finally, $\mathbb{J}_k = \mathbb{J}_k^{(4)} \cup \mathbb{J}_k^{(3)} \cup \mathbb{J}_k^{(2)} \cup \mathbb{J}_k^{(1)} \cup \mathbb{J}_k^{(0)}$ represents the user set that potentially interferes with user k. Note that the classification of the above five user sets is determined by the network topology. Figure 4.2 shows an illustrative diagram of the neighboring user classification.

Denote the available channel set as $\mathbb{M}$, i.e., $\mathbb{M} = \{1, 2, \ldots, M\}$. Assume that user k chooses channel a_k, and the channel separation between users k and j is $\delta(a_k, a_j)$. Then, the number of interfering users in the five potentially interfering sets, i.e., $\mathbb{J}_k^{(i)}$, $\forall i = 0, 1, \ldots, 4$, can be calculated as

$$s_k^{(i)} = \sum_{j \in \mathbb{J}_k^{(i)}} \sigma_{kj}^{(i)}, \tag{4.5}$$

where $\sigma_{kj}^{(i)}$ is the MAC-layer interference between users k and j in region $\mathbb{J}_k^{(i)}$, i.e.,

$$\sigma_{kj}^{(i)} = \begin{cases} 1, & \delta(a_k, a_j) \leq i \\ 0, & \text{otherwise} \end{cases} \tag{4.6}$$

Fig. 4.2 An illustrative
diagram of the neighboring
user classification

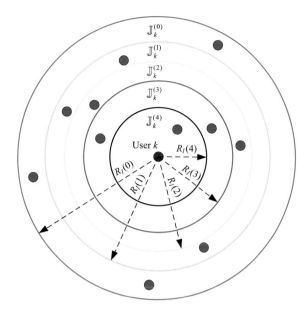

Thus, according to the principle of the contention coordination mechanisms in
802.11b-based networks, the achieved normalized throughput of user k under a channel selection profile $(a_1, \ldots, a_K)$ is given by

$$T_k(a_1, \ldots, a_K) = \frac{1}{1 + s_k}, \tag{4.7}$$

where s_k is aggregate MAC-layer interference experienced by user k in range $\mathbb{J}_k$, i.e.,

$$s_k = \sum_{i=0}^{4} \sum_{j \in \mathbb{J}_k^{(i)}} \sigma_{kj}^{(i)} = \sum_{i=0}^{4} s_k^{(i)}. \tag{4.8}$$

Following the similar methodology in Chap. 3, we formulate an interference mitigation problem with partially overlapping channels. Specifically, the optimization
objective is to find the optimal channel selection profile to minimize the aggregate
MAC-layer interference of all the users in the network, i.e.,

$$\textbf{P1:} \quad \min \sum_{k \in \mathbb{K}} s_k \tag{4.9}$$

However, **P1** is a combinatorial problem which is hard to solve even in a centralized manner. Thus, the task of solving **P1** in a distributed manner is challenging. In
the following, an efficient game-theoretic learning approach is proposed.

4.3 Graphical Game Model

4.3.1 Graphical Game Model

Formally, we formulate the interference mitigation problem with partially overlapping channels as a non-cooperative game. Specifically, the game is denoted as
$G = \{\mathbb{K}, A_k, \mathbb{J}_k, u_k\}$, where $\mathbb{K}$ is the player (user) set, A_k is the action (available
channel) set of player k, $\mathbb{J}_k$ in the neighboring user set of player k, and u_k is the utility function. Note that all channels are available for the each player, which means that
the action set of all the players is exactly the channel set, i.e., $A_k \equiv \mathbb{M}, \forall k \in \mathbb{K}$. In
the considered scenarios, the utility function of an arbitrary player k is only directly
affected by its action and the action profile of the players in $\mathbb{J}_k$. Thus, the formulated game model belongs to local interactive game [3] or graphical game [16]. To
capture such local interaction, the utility function of player k can be expressed as
$u_k(a_k, a_{\mathbb{J}_k})$, where a_k is the channel selection of player k and $a_{\mathbb{J}_k}$ is the action profile
of the players in $\mathbb{J}_k$. In the formulated interference mitigation game with partially
overlapping channels, the utility function is defined as

$$u_k(a_k, a_{\mathbb{J}_k}) = -s_k, \tag{4.10}$$

where s_k is the MAC-layer interference experienced by player k, as defined in (4.8).
Then, it is expressed as follows:

$$G: \quad \max u_k(a_k, a_{\mathbb{J}_k}), \forall k \in \mathcal{K} \tag{4.11}$$

4.3.2 Analysis of Nash Equilibrium

In this following, the properties of the formulated interference mitigation game in
terms of Nash equilibrium (NE) existence and achievable performance are analyzed.

Theorem 4.1 *The interference mitigation game with partially overlapping channels
is an exact potential game which has at least one pure strategy NE point.*

Proof According to the definition of exact potential game (see Definition 1.2 in
Chap. 1), we need to prove that there exists a potential function such that the change
in the individual utility function caused by any player's unilateral deviation is the
same as that in the potential function. Specifically, we define the potential function
as follows:

$$\Phi(a_1, \ldots, a_K) = -\frac{1}{2} \sum_{k \in \mathbb{K}} s_k, \tag{4.12}$$

which is exactly the negative value of the half aggregate MAC-layer interference. Then, it can be verified that it is an exact potential game with Φ serving as the potential function. For detailed lines for proof, refer to [15]. □

As the players selfishly maximize their individual utility functions, as specified by (4.11), it may lead to inefficiency and dilemma, which is known as *tragedy of commons* [17]. Therefore, it need to investigate the achievable performance of NE solutions. To begin with, the aggregate MAC-layer interference of a pure strategy NE point $a_{\text{NE}} = \{a_1^*, \ldots, a_N^*\}$ is given by

$$U(a_{\text{NE}}) = \sum_{n \in \mathcal{K}} s_k(a_k^*, a_{\mathcal{J}_k}^*), \tag{4.13}$$

where s_k is calculated by (4.8). Then the achievable performance of NE solutions of the formulated game is characterized by the following theorems.

Theorem 4.2 *The global minimum of the aggregate MAC-layer interference of the network constitutes a pure strategy NE point of G.*

Proof Suppose that a_{opt} is an optimal channel selection profile that maximizes the potential function, i.e.,

$$a_{opt} \in \arg\max_a \Phi(a_1, \ldots, a_K). \tag{4.14}$$

Using the relationship between the potential function and the optimization objective, as specified by (4.12), a_{opt} also minimizes the aggregate MAC-layer interference in the network, i.e.,

$$a_{opt} \in \arg\min_a \sum_{k \in \mathcal{K}} s_k \tag{4.15}$$

Since a maximizer of the potential function constitutes a pure strategy NE point of any exact potential game [18], it follows that a_{opt} serves as the optimal pure strategy NE of formulated MAC-layer interference mitigation game. □

Theorem 4.3 *For any network topology, the aggregate MAC-layer interference is upper bounded by* $U(a_{NE}) < \sum_{k \in \mathbb{K}} ((\sum_{i=0}^{4}(2i + 1)|\mathbb{J}_k^{(i)}|)/M)$.

Proof Refer to [15]. □

Interestingly, it can be observed from Theorem 4.3 that increasing the total number of available channels, i.e., M, would lower the aggregate MAC-layer interference in the network. This is exactly the basic idea of using partially overlapping channels instead of orthogonal channels.

4.4 Simultaneous Log-Linear Learning Algorithm with Heterogeneous Rates

4.4.1 Algorithm Description

As stated before, there are a large number of learning algorithms that can achieve pure strategy Nash equilibria of potential games, e.g., the best response [12], the stochastic learning automata used in Chap. 2, and the binary log-linear learning algorithm used in Chap. 3. In the stochastic automata learning algorithm, all the players can perform learning simultaneously but the convergent NE solutions are not optional in general scenarios. Although the binary log-linear can achieve the best Nash equilibrium, there is a restriction that only one player is allowed to update its action at a time. Note that it needs some distributed coordination mechanisms to schedule exactly one player to learn in each iteration, and the most efficient approach is to use a common control channel (CCC) for coordination. However, a CCC is not always available in practice. Thus, it needs to develop a simultaneous and uncoupled algorithm that can achieve the optimal solution.

As the MAC-layer interference cannot be observed directly by the users, a simple and efficient estimation approach is then proposed. Suppose that there are N slots in each iteration of the learning algorithm, and denote the number of slots in which player k successfully access the channel in the ith iteration as $N_k(i)$. Then, the expected achievable throughput of player k can be expressed as $T_k = \frac{N_k(i)}{N} = \frac{1}{1+s_k(i)}$ and hence the aggregate interference experienced by player k can be estimated as $\hat{s}_k(i) = \frac{N}{N_k(i)} - 1$, which means that the received utility in each iteration can be estimated by

$$\hat{u}_k(i) = 1 - \frac{N}{N_k(i)}, \tag{4.16}$$

In this chapter, a simultaneous log-linear learning algorithm is proposed, in which all the players simultaneously update their actions in each play. The procedure of the proposed learning algorithm is described in Algorithm 3. The key idea can be explained as follows: (i) in each iteration, the players explore new actions with probability δ_k, as given in (4.17). (ii) After the exploration, they update their actions using the log-linear learning rule over the actions in the last two iterations, as shown in (4.18). An illustrative diagram of the simultaneous log-linear learning algorithm is shown in Fig. 4.3. It is noted that the proposed learning algorithm is simultaneous, fully distributed, and uncoupled.

4.4.2 Convergence Analysis

In methodology, the simultaneous log-linear learning algorithm is motivated by the spatial adaptive play (SAP) [3, 19, 20], which is a coupled learning algorithm for that

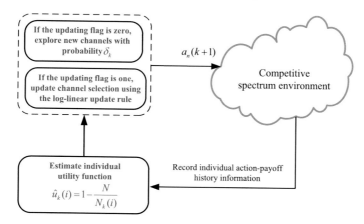

Fig. 4.3 The illustrative diagram of the simultaneous log-linear learning algorithm

---

Algorithm 3: simultaneous log-linear learning algorithm

Initialization: For iteration index $i = 0$, each player k, $\forall k \in \mathbb{K}$, randomly selects a channel $a_k(0) \in A_k$. An updating binary flag, $x_k(0) = 0$, $\forall k \in \mathbb{K}$, is set for each player k.
All players simultaneously execute the following procedure:
Loop for $i = 1, \ldots,$
Exploration:
If the updating flag is zero, i.e., $x_k(i-1) = 0$, player k explores all the possible channels using to the following rule:

$$\Pr[a_k(i) = a] = \begin{cases} \frac{\delta_k}{|A_k|-1}, & a \in \{A_k \backslash a_k(i-1)\} \\ 1 - \delta_k, & a = a_k(i-1), \end{cases} \tag{4.17}$$

where $|A_k|$ is the number of available channels of player k, and δ_k can be regarded as the exploration rate. Furthermore, set $x_k(i) = 1$ if $a_k(i) \neq a_k(i-1)$, and $x_k(i) = 0$ otherwise.
End if
Update:
If the updating flag is one, i.e., $x_k(i-1) = 1$, player k updates the action according to the following rule:

$$\Pr[a_k(i) = a_k(i-1)] = \frac{\exp\{\hat{u}_k(i-1)\beta\}}{\exp\{\hat{u}_k(i-1)\beta\} + \exp\{\hat{u}_k(i-2)\beta\}}$$

$$\Pr[a_k(i) = a_k(i-2)] = \frac{\exp\{\hat{u}_k(i-2)\beta\}}{\exp\{\hat{u}_k(i-1)\beta\} + \exp\{\hat{u}_k(i-2)\beta\}}, \tag{4.18}$$

where β is the learning parameter, $\hat{u}_k(i-1)$ and $\hat{u}_k(i-2)$ represent the received utility function of the player k in iterations $i-1$ and $i-2$ respectively, as specified by (4.16). Set $x_k(i) = 0$.
End if
End loop

---

requires information exchange among the users. Although there is an improved version of SAP, which is called the binary log-linear algorithm, no information exchange
is needed. (This algorithm was used in Chap. 3). However, only one player is allowed
to perform learning in each iteration. In essence, the used simultaneous log-linear
learning algorithm is a variant of the payoff-based learning algorithm which was originally proposed in [21]. The key difference is that homogeneous exploration rate was
used in [21] while heterogeneous exploration rates are used in the proposed learning algorithm. Specifically, the following theorem characterizes the convergence and
optimality of the simultaneous log-linear learning algorithm.

Theorem 4.4 *If the exploration parameters are chosen as* $\delta_k = \exp(-\beta m_k)$,
$\forall k \in \mathcal{K}$, *the simultaneous log-linear learning algorithm asymptotically converges
to an optimal channel selection profile that minimizes the aggregate MAC-layer
interference for sufficiently large* m_k.

Proof The proof is mainly based on the theory of resistance trees [21, 22] and detailed
lines can be found in [15].

4.4.2.1 Discussion on the Learning Parameters

In this part, we discuss the rationale behind the learning algorithm. At the beginning,
the players randomly explore new actions for finding a better channel. Based on the
results of exploration, the log-linear updating strategy is used to update its selection.
Note that the log-linear strategy is also known as Boltzmann exploration strategy
[23], in which the probability of choosing an action with higher utility is larger than
that for an action with lower utility. The Boltzmann rule is an efficient way to escape
from local optimal points and finally achieves the global optimum. Thus, in order to
balance the tradeoff between exploration and convergence speed, it is desirable to set
the value of the learning parameter m_k small values at the beginning phase and let it
increase as the algorithm iterates. In practice, the following simplest linear strategy
can be used: $m(i) = m_0 + i \Delta m$, where m_0 is the initial value, Δm is the step size,
and i is the iteration number.

The simultaneous log-linear learning algorithm was originally proposed in [21],
in which globally interactive games, i.e., an action of a player affects all other players,
were considered. In the local interactive games (graphical games), i.e., games over
graph, we found that exploiting the feature of the spatially locations of the players
would accelerate the learning speed, which motivates us to set the heterogeneous
exploration rates for the players. Denote the number of potentially interfering users
of player k, i.e., the neighboring players, as $D_k = |\mathbb{J}_k|$. To achieve heterogeneous
exportation rates, the players with less value of D_k are advisable to have larger
exploration rates while those with large value of D_k are advisable to have smaller
rates. The reason is as follows: the players with large value of D_k, i.e., they have
more neighboring users, have more impact on the system, and hence their action
changes would result in more perturbations and lower convergence speed. Thus, we

set the exploration rate of player k as $m_k(i) = \frac{\max\{D_1,...,D_k\}}{D_k} m(i)$. It will be shown later that heterogeneous exploration rates lead to faster convergence speed than that of homogeneous exploration rates.

4.5 Simulation Results and Discussion

4.5.1 Scenario Setup

We consider IEEE 802.11b-based networks with 2 Mb/s data rate. All the users are located in a $1000\,m \times 1000\,m$ square area, and the interference range of co-channel communication is 200 m [8], i.e., $2R = 200\,m$. Thus, according to Table 4.1, the interference ranges of different channel separations as follows: $R_I(1) = 112.5\,m$, $R_I(2) = 75\,m$, $R_I(3) = 37.5\,m$, and $R_I(4) = 12.5\,m$. To make it more general, two kinds of networks are considered. The first is the random topology, in which the users are randomly located in the square, and the second is the grid topology, in which the users are located in the junction points of a grid. In order to investigate the impact of the user density, the number of users is arbitrarily increased in the random topology. For the grid topology, the number of users is increased in a square rule, i.e., the number of users is given by $K = l^2$, where l is a natural number. Examples of the considered random and grid topologies are given by Figs. 4.4 and 4.5 respectively.

To evaluate the throughput improvement of partially overlapping channels (POC) over non-overlapping channels (NOC), we compare the achievable throughput of the proposed POC approach with that of an existing optimal NOC approach using the spatial adaptive play (SAP) approach [3]. In that approach, the MAC-layer interference mitigation problem was also formulated as an exact potential game and the optimal optimum is achieved via information exchange. In addition, since the

Fig. 4.4 An example of the random topology with 60 users. The *small solid circles* represent the users, while the large *dashed circles* represent the interference regions corresponding to different channel separations

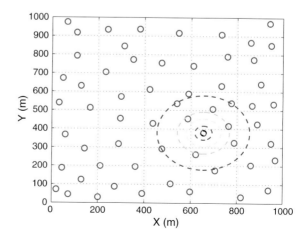

Fig. 4.5 An example of the grid topology with 64 users. The *small solid circles* represent the users, while the large *dashed circles* represent the interference regions corresponding to different channel separations

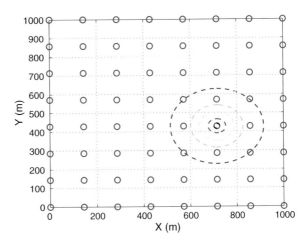

formulated interference mitigation problem with POC is an exact potential game with the optimal NE solutions minimizing the aggregate MAC-layer interference, other learning algorithms that can converge to the pure strategy NE solutions can also be used as referred algorithms. Thus, we also evaluate the throughput performance of the SAP algorithm [3], the MAX log-linear algorithm [24], the binary log-linear algorithm (B-logit) [25], and the multi-agent Q-learning algorithm[1] [26]. In classification, SAP is coupled and sequential (only one player updates action at a time), both MAX-logit and B-logit are uncoupled and sequential, and both the simultaneous log-linear algorithm and the Q-learning algorithm are uncoupled and simultaneous (multiple players update actions simultaneously).

4.5.2 Convergence Behavior

In this part, the convergence behaviors of the simultaneous log-linear learning algorithm are studied. The learning is set to $\beta = -8$ and $m = 0.1 + 0.0095i$, where i is the iteration number. Note that these parameters have been optimized by experiments.

First, we present the convergence behaviors of the proposed learning algorithm with partially overlapping channels (POC). The convergence behaviors for the above two example networks are shown in Fig. 4.6. These results are obtained by taking the expected value of 20 independent trials. It is noted that for both topologies, the proposed learning algorithm converges in about 600 iterations. Also, it is noted from the figure that, as the algorithm iterates, the aggregate interference in both topologies

---

[1] We found that it never converges when the original Q-learning algorithm presented in [26] is directly applied. To achieve its convergence, we modify the algorithm slightly by setting the learning parameter as $\gamma = \frac{1}{5+0.045i}$, where i is the iteration number.

Fig. 4.6 The convergence behaviors of the proposed learning algorithm using partially overlapping channels for the two example topologies

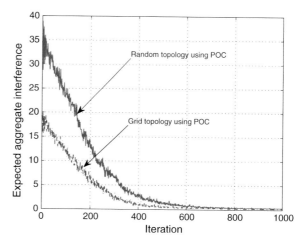

Fig. 4.7 The convergence behaviors of the proposed learning algorithm using non-overlapping (orthogonal) channels for the two example topologies

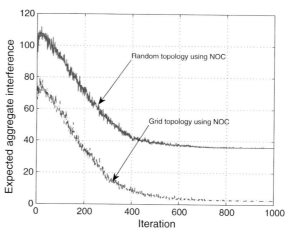

decreases to zero gradually. These results validate the convergence of the proposed synchronous learning algorithm with partially overlapping channels.

Second, we study the convergence behaviors of the proposed learning algorithm with non-overlapping channels (NOC). The convergence behaviors for the two example network topologies are shown in Fig. 4.7. It is noted that the algorithm in both topologies also converges in about 600 iterations. These results validate the generality of the proposed learning algorithm, since it is not only suitable for partially overlapping channels but also suitable for non-overlapping channels. In addition, it is shown that the final aggregate interference using NOC is much greater than those using POC for both network topologies. According to the relationship between the experienced interference and the achieved throughput, it can be inferred that the throughput performance of POC is much higher than that of NOC.

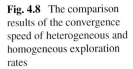

Fig. 4.8 The comparison results of the convergence speed of heterogeneous and homogeneous exploration rates

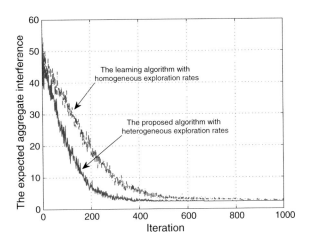

Third, the convergence comparisons of homogeneous and heterogeneous exploration rates are shown in Fig. 4.8, in which the results are obtained by simulating 20 independent trials and then taking the expected value. It is observed that the learning algorithm with heterogeneous exploration rates converges in about 400 iterations while the algorithm with homogeneous rates converges in about 600 iterations. Also, the aggregate interference of both algorithms decreases gradually. These results validate the faster convergence speed caused of heterogeneous exploration rates.

4.5.3 Performance Evaluation

In this part, the achieved performance of the simultaneous log-linear algorithm using POC is evaluated. In particular, the simultaneous log-linear algorithm is compared with other coupled and uncoupled algorithms, and the achieved performance of POC is also compared that of NOC.

4.5.3.1 Random Topology

First, the comparison results of the expected aggregate interference of random topologies are shown in Fig. 4.9. The number of users increases from 40 to 150. The results are obtained by independently simulating 500 topologies and then taking the expected values. It is noted that the expected aggregate interference of POC is greatly less than that of NOC. In particular, as the number of users becomes larger, the differences become larger, as can be expected in any wireless networks.

Second, the comparison results of the expected network throughput of random topologies are shown in Fig. 4.10. It is noted that the expected network throughput

Fig. 4.9 The comparison results of the expected aggregate interference of random topologies

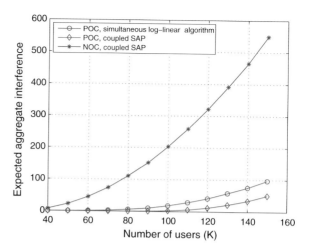

Fig. 4.10 The comparison results of the expected network throughput of random topologies

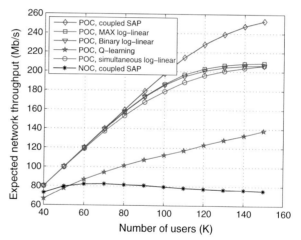

of POC is significantly larger than that of NOC, which validates the statement that the throughput improvement of POC over NOC is significant. Also, it is observed that the simultaneous log-linear algorithm achieves satisfactory performance when compared with other existing learning algorithms. In particular, for relatively sparse networks, e.g., $K \leq 80$, the achieve throughput of the simultaneous log-linear algorithm is almost the same with that of SAP. As the user density increases, it is slightly worse than that of SAP. For other uncoupled algorithms, it can be observed that the throughput performance of the simultaneous log-linear algorithm is close to those of MAX log-linear and binary log-linear algorithms, which belong to uncoupled and sequential algorithms. In addition, the simultaneous log-linear algorithm achieves much higher throughput than that of Q-learning, which is also uncoupled and synchronous.

4.5.3.2 Grid Topology

First, the comparison results of the expected aggregate interference of grid topologies are shown in Fig. 4.11. The number of nodes increases from 36 to 169. The results are obtained by independently running 500 trials and then taking the expected values. It is noted that the expected interference of POC is significantly less than that of NOC, which exhibits the same trend as that in random topologies.

Second, the comparison results of the expected network throughput of grid topologies are shown in Fig. 4.12. It is also observed that the expected network throughput of POC is much higher than that of NOC, especially when the number of users is large. For scenarios with small number of users, e.g., $K = 36, 49, 64, 81$, the throughput gap is trivial. The reason is that the interference in the grid topologies is light in sparse networks. An interesting result in Fig. 4.12 is that the expected network throughput

Fig. 4.11 The comparison results of the expected aggregate interference of grid topologies

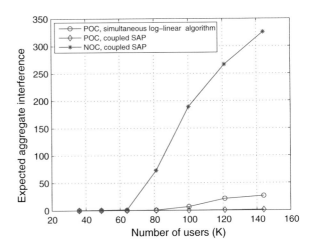

Fig. 4.12 The comparison results of the expected network throughput of grid topologies

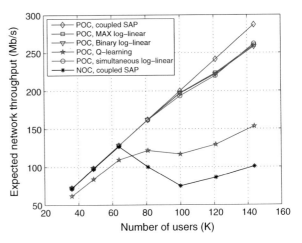

of NOC exhibits a singular trend. In particular, it increases when the number of users K increases from 36 to 64, decreases when K increases from 64 to 100, and again increases when K increases from 100 to 144. The potential reason might be as follows: as the users in the grid networks are fixed, some scales of the networks may lead to singular results. Furthermore, the simultaneous log-linear learning algorithm achieves satisfactory performance when compared with other existing coupled or uncoupled algorithms, and the same trend as that in random topologies also holds.

4.5.3.3 Throughput Gain of POC Over NOC

To characterize the throughput gain of POC over NOC, we present the throughput gain for the two kinds of network topologies in Fig. 4.13. The throughput gain is defined as the incremental ratio of the expected network throughput POC and NOC, respectively, i.e., it is expressed as

$$g = \frac{T_{poc}}{T_{noc}} - 1, \tag{4.19}$$

where T_{poc} is the achieved throughput of using POC and T_{noc} is that of using NOC. It is seen from the figure that when the number of nodes increases from 40 to 150, the throughput gain of random topologies increases from 15 to 180 %. Also, for the grid topologies, the throughput gain is trial (about 2 %) when the number of nodes is less than 80, and increases from 70 to 140 % when the number of nodes increases from 80 to 140. A common trend for the two kinds of network topologies is that the throughput gain is small in sparse networks, i.e., the number of nodes is small, and becomes larger in more dense networks, i.e., the networks with more nodes. This can be explained as follows: (i) the interference in sparse networks is slight and that

Fig. 4.13 The comparison results of throughput gains of the proposed uncoupled learning algorithm

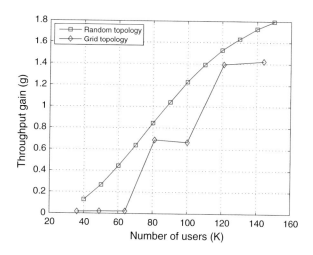

in dense networks is heavy, and (ii) for sparse networks, NOC provides with enough channels and hence achieves the same throughput with POC.

4.6 Concluding Remarks

Compared with orthogonal (non-overlapping) channels, as the one considered in Chap. 3, partially overlapping channels improve the full-range spectrum utilization significantly. The key design is that the channel separation and physical distance are jointly considered in the interference model. To meet the increasing mobile traffic in future wireless networks, exploiting partially overlapping channels is a promising candidate. For example, one may introduce partially overlapping channels into the LTE-U systems, in which the users dynamically share the 802.11a bands with WLAN.

References

1. M. Felegyhazi, M. Cagalj, J.P. Hubaux, Efficient MAC in cognitive radio systems: a game-theoretic approach. IEEE Trans. Wirel. Commun. **8**(4), 1984–1995 (2009)
2. M. Maskery, V. Krishnamurthy, Q. Zhao, Decentralized dynamic spectrum access for cognitive radios: cooperative design of a non-cooperative game. IEEE Trans. Commun. **57**(2), 459–469 (2009)
3. Y. Xu, J. Wang, Q. Wu et al., Opportunistic spectrum access in cognitive radio networks: global optimization using local interaction games. IEEE J. Sel. Signal Process **6**(2), 180–194 (2012)
4. H. Li, Z. Han, Competitive spectrum access in cognitive radio networks: graphical game and learning, in *Proceedings of IEEE WCNC* (2010), pp. 1–6
5. Y. Xu, Q. Wu, J. Wang et al., Distributed channel selection in CRAHNs with heterogeneous spectrum opportunities: a local congestion game approach. IEICE Trans. Commun. **E95–B**(3), 991–994 (2012)
6. A. Mishra, E. Rozner, S. Banerjee, W. Arbaugh, Exploiting partially overlapping channels in wireless networks: turning a peril into an advantage, in *Proceedings of ACM SIGCOMM* (2005), pp. 29–29
7. A. Mishra, S. Banerjee, W. Arbaugh, Weighted coloring based channel assignment for WLANs. ACM SIGMOBILE Mob. Comput. Commun. Rev. **3**, 19–31 (2005)
8. Z. Feng, Y. Yang, How much improvement can we get from partially overlapped channels? in *Proceedings of IEEE WCNC* (2008), pp. 2957–2962
9. Y. Ding, Y. Huang, G. Zeng, L. Xiao, Channel assignment with partially overlapping channels in wireless mesh networks, in *WICON* (2008)
10. M. Hoque, X. Hong, F. Afroz, Multiple radio channel assignement utilizing partially overlapped channels, in *Proceedings of IEEE GLOBECOM'09* (2009), pp. 1–7
11. P. Duarte, Z. Fadlullah, K. Hashimoto, N. Kato, Partially overlapped channel assignment on wireless mesh network backbone, in *IEEE Proceedings of Globecom'10* (2010)
12. P. Duarte, Z. Fadlullah, A. Vasilakos, N. Kato, On the partially overlapped channel assignment on wireless mesh network backbone: a game theoretic approach. IEEE J. Sel. Areas Commun. **30**(1), 119–127 (2012)
13. Y. Ding, Y. Huang, G. Zeng, L. Xiao, Using partially overlapping channels to improve throughput in wireless mesh networks, IEEE Trans. Mob. Comput. (to appear)
14. Y. Cui, W. Li, X. Cheng, B. Chen, Partially overlapping channel assignment based on "node orthogonality" for 802.11 wireless networks, IEEE Trans. Mob. Comput. (to appear)

15. Y. Xu, Q. Wu, J. Wang, L. Shen, A. Anpalgan, Opportunistic spectrum access using partially overlapping channels: graphical game and uncoupled learning. IEEE Trans. Commun. **61**(9), 3906–3918 (2013)
16. C. Tekin, M. Liu, R. Southwell, J. Huang, S. Ahmad, Atomic congestion games on graphs and their applications in networking, IEEE/ACM Trans. Netw. **20**(5) (2012)
17. H. Kameda, E. Altman, Inefficient noncooperation in networking games of common-pool resources. IEEE J. Sel. Areas Commun. **26**(7), 1260–1268 (2008)
18. D. Monderer, L.S. Shapley, Potential games. Games Econ. Behav. **14**, 124–143 (1996)
19. H.P. Young, *Individual Strategy and Social Structure* (Princeton University Press, Princeton, 1998)
20. Y. Xu, Q. Wu, J. Wang, Y.-D. Yao, Social welfare maximization for SRSNs using bio-inspired community cooperative mechanism. Chin. Sci. Bull. **57**(1), 125–131 (2012)
21. J. Marden, J. Shamma, Revisiting log-linear learning: asynchrony, completeness and payoff-based implementation. Games Econ. Behav. **75**(2), 788–808 (2012)
22. H.P. Young, The evolution of conventions. Econometrica **61**(1), 57–84 (1993)
23. R.S. Sutton, A.G. Barto, *Reinforcement Learning: An Introduction* (MIT Press, Cambridge, 1998)
24. Y. Song, S. Wong, K.-W., Lee, Optimal gateway selection in multi-domain wireless networks: a potential game perspective, in *Proceedings of ACM MobiCom* (2011)
25. J. Marden, G. Arslan, J. Shamma, Cooperative control and potential games. IEEE Trans. Syst. Man Cybern. B **39**(6), 1393–1407 (2009)
26. H. Li, Multi-agent Q-learning for Aloha-like spectrum access in cognitive radio systems, EURASIP J. Wirel. Commun. Netw. **2010**, 1–15 (2010)

Chapter 5
Robust Interference Coordination with Dynamic Active User Set

5.1 Introduction

Due to hardware limitation, the users in dynamic spectrum access networks can sense only a small part of channels (always one) at a time [1]. As a result, there are two basic channel sensing strategies [2]: parallel sensing, i.e., a fixed set of channels is simultaneously sensed in each slot, and sequential sensing, i.e., channels are sequentially sensed according to a pre-defined order. For parallel sensing strategies, the users have to keep silent in the current slot if no idle channel is found, which may be inefficient. In comparison, the sequential sensing is more efficient and adaptive. However, interference/collision occurs if more than two users sense and access an idle channel simultaneously. Thus, the sensing orders in the sequential sensing strategy should be carefully designed [3–6]. In Chaps. 2, 3, and 4, parallel sensing strategies were considered. In this chapter, we consider the problem of interference coordination for sequential dynamic spectrum access, and the focus is to optimize channel sensing orders of the users.

The problem of channel sensing order optimization for single-user dynamic spectrum access systems has been extensively studied in the literature [7–13]. For multi-user dynamic spectrum access networks, it has begun to draw attention very recently [3–6]. Note that in most existing decision-making optimization problems for dynamic spectrum access networks, the number of active users is assumed fixed. However, it should be pointed out that in several practical scenarios, a user does not perform learning when there is no traffic to transmit. Thus, it is important and timely to investigate the impact of dynamic active users on game formulation and learning algorithm.

In this chapter, a generalized interference metric is proposed to address the overlapping of multiple channel orders. Based on this, the following two optimization objectives are considered: minimizing the aggregate interference for a specific active user set and minimizing the expected aggregate interference for over all possible active user sets. To achieve distributed decision making, two interference mitigation

© The Author(s) 2016
Y. Xu and A. Anpalagan, *Game-theoretic Interference Coordination Approaches for Dynamic Spectrum Access*, SpringerBriefs in Electrical and Computer Engineering, DOI 10.1007/978-981-10-0024-9_5

game models are proposed: (i) a state-based game, in which the active user set characterizes the system state in each slot, and (ii) a robust game, in which the utility functions are defined as the expected value over all system states. The two game models are proved to be exact potential games, and the stochastic learning automata algorithm is applied to achieve stable and desirable solutions. Note that some useful tutorials for robust games with changing player set can be found in [14], and the main analysis and results in this chapter were presented in [15].

5.2 System Model and Problem Formulation

5.2.1 System Model

Consider a distributed dynamic spectrum access network with N secondary users and M channels. Assume that time is divided into slots with equal length, and denote x_m as the occupancy state of channel m. Specifically, $x_m = 1$ means that channel m is idle and $x_m = 0$ means it is unavailable. For analysis simplicity, assume that the states of the channels are determined by the idle channel probabilities, θ_m, $0 \leq \theta_m \leq 1, \forall m \in \{1, 2, \ldots, M\}$, and remain unchanged in a slot. When the number of users is larger than that of channels, i.e., $N > M$, the users would choose a fixed number of channels (always one) to sense and access in a slot rather than performing the sequential sensing strategies, as the spectrum opportunities are limited in this scenario. Thus, we only consider the case that the number of the users is not greater than that of the licensed channels, i.e., $N \leq M$ in this chapter.

To make it more practical, dynamic active user model is considered. Specifically, each user performs channel sensing and accesses in each slot with a probability λ_n, $0 < \lambda_n \leq 1$. Note that such a model captures general kind of dynamics in dynamic spectrum access networks, e.g., a user becomes active only when there is data to transmit, a mobile user joins or leaves dynamically. Furthermore, the active probability can be regarded as the probability of non-empty buffer.

5.2.2 Problem Formulation

For presentation and analysis, a system state is defined as $\mathbf{S} = \{s_1, \ldots, s_N\}$, where $s_n = 1$ means that the nth user is active, while $s_n = 0$ indicates that it is inactive. Then, the probability of a system state can be expressed as $\mu(s_1, \ldots, s_N) = \prod_{n=1}^{N} p_n$, where p_n is determined as follows:

$$p_n = \begin{cases} \lambda_n, & s_n = 1 \\ 1 - \lambda_n, & s_n = 0 \end{cases} \tag{5.1}$$

Denote the user set in the network as $\mathbb{N}$, i.e., $\mathbb{N} = \{1, \ldots, N\}$, the active user set as $\mathbb{B}$, i.e., $\mathbb{B} = \{n \in \mathbb{N} : s_n = 1\}$, and the set of all possible active user sets as Γ. Then, the probability of an specific active user set is $\mu(\mathbb{B})$, which satisfies $\sum_{\mathbb{B} \in \Gamma} \mu(\mathbb{B}) = 1$. However, it should be pointed out that the distribution probabilities of the system states are unknown to the users since there is no information exchange among the users.

The impact of the channel sensing orders of the users on the achievable network throughput is analyzed by the following examples. Consider a dynamic spectrum access system with five channels and two active users. Their channel sensing orders are $\{1, 3, 2, 4, 5\}$ and $\{3, 4, 2, 5, 1\}$, respectively, which means that the two users simultaneously sense channel 2 at time 3τ. Then, if channel 2 is detected as idle by both users, they may access the channel simultaneously and hence cause collision. An illustrative diagram of the example scenario is shown in Fig. 5.1. Note that the two users can employ some approaches, e.g., cognitive TDMA and CSMA, to resolve collision.

For presentation, denote the permutation set of M as $\mathbb{O}$, and the channel sensing order of the nth user as a M-dimensional order vector $O_n = (o_{n1}, o_{n2}, \ldots, o_{nM})$, $O_n \in \mathbb{O}$. Motivated by the MAC-layer interference model analyzed in Chaps. 3 and 4, a more general interference metric can be defined to capture the impact of accessing the same channel simultaneously. Specifically, the generalized interference metric between two active users n and m is defined as follows:

$$g_{nm} = O_n \odot O_m, \tag{5.2}$$

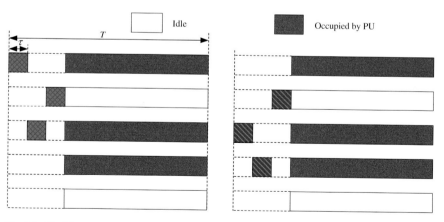

Fig. 5.1 The illustrative diagram of a system with five channels and two users. The channel sensing order of user 1 is $\{1, 3, 2, 4, 5\}$ and that of user 2 is $\{3, 4, 2, 5, 1\}$. When detecting channel 2 as idle in time 3τ, they access this channel simultaneously and cause collision

where $\odot$ is the bitwise XNOR operation. That is, $g_{nm} = \sum_{k=1}^{M} \delta(o_{nk}, o_{mk})$, and $\delta(o_{nk}, o_{mk})$ is the following indicator function:

$$\delta(o_{nk}, o_{mk}) = \begin{cases} 1, & o_{nk} = o_{mk} \\ 0, & o_{nk} \neq o_{mk} \end{cases} \tag{5.3}$$

If the interference between users n and m is zero, i.e., $g_{nm} = 0$, we say their channel sensing orders are orthogonal. It can be seen that g_{nm} reflects the impact of overlapped channel sensing orders on the achievable throughput. Specifically, larger value of g_{nm} causes lower achievable throughput, and vice versa. Extending the interference model to the scenarios of multiple users, the aggregate generalized interference level experienced by an active user n in a slot is defined as follows:

$$G_n(\mathbb{B}) = \sum_{m \in \mathbb{B}, m \neq n} g_{nm} = \sum_{m \in \mathbb{B}, m \neq n} O_n \odot O_m \tag{5.4}$$

A lower value of $G_n(\mathbb{B})$ is desirable for user n as higher throughput can be achieved. Similarly, a lower value of an aggregate generalized interference level of all the users is also desirable. Thus, we define the aggregate generalized interference level of an active user set $\mathbb{B}(k)$ as follows:

$$I(\mathbb{B}) = \sum_{n \in \mathbb{B}} \sum_{m \in \mathbb{B}, m \neq n} O_n \odot O_m \tag{5.5}$$

If $I(\mathbb{B}) = 0$, it corresponds to an interference-free profile of channel sensing orders as there is no overlap in the channel sensing orders of any two users. Formally, the optimization objective is to find an optimal channel sensing order profile to maximize the aggregate interference level, i.e.,

$$\textbf{(P1:)} \qquad \max \; -I(\mathbb{B}), \forall \mathbb{B} \in \Gamma \tag{5.6}$$

or

$$\textbf{(P2:)} \qquad \max \; -\mathrm{E}_{\mathbb{B}}[I(\mathbb{B})] = \sum_{\mathbb{B} \in \Gamma} \mu(\mathbb{B})I_{\mathbb{B}}, \tag{5.7}$$

where $\mathrm{E}_{\mathbb{B}}[\cdot]$ takes the expectation over all possible active user sets. Generally, solving problem **P1** and **P2** in a distributed manner is challenging, and the reasons are as follows: (i) they are combinatorial optimization problems, which are NP-hard, and (ii) the active user set in each slot is unknown to the users, since a user only knows its own state but knows nothing about other users. In the following, a game-theoretic distributed solution is developed for solving the two problems.

Remark 5.1 The optimality of the above optimization problems is discussed as follows. For a network with homogeneous channels, the optimality in interference means the optimality in throughput, on the condition that the final result is interference-free. For a network with heterogeneous channels, this is not true. However, we would like to point out that due to the heavy computational complexity, it is not possible to achieve an optimal solution for sum-rate maximization directly. For example, considering a system with five channels and five users, the total possible solutions are $(5!)^5$, which is extremely huge. Thus, although the sum-rate maximization generally cannot be achieved, we believe that the formulated interference mitigation approaches are desirable as distributed solutions with low-complexity can be achieved.

5.3 Channel Sensing Order Selection Games

In this section, we formulate two game non-cooperative models to address the formulated channel sensing order optimization problems **P1** and **P2**, respectively. The first one is a state-based order selection game, in which a system state is defined to describe the active user set in each slot, and the second one is a robust order selection game, in which the utility functions are defined as the expectation value over all possible system states. Their properties in terms of existence of Nash equilibria and achievable performance are analyzed.

5.3.1 State-Based Order Selection Game

5.3.1.1 Game Model

In this game, a system state is added to capture the dynamic active user set. Specifically, the state-based game is denoted as $\mathbb{G}_1 = \{\mathbb{N}, \mathbb{B}, \{A_n\}_{n \in \mathbb{B}}, \{U1_n\}_{n \in \mathbb{B}}\}$, where $\mathbb{N}$ is the potential player (user) set, i.e., $\mathbb{N} = \{1, 2, \ldots, N\}$, $\mathbb{B}$ is the system state which corresponds to the active user set in the current slot, A_n is the available action set of player n, and $U1_n$ is the utility function of player n. Note that all the players' action sets are the permutation set of M, i.e., $\mathbb{A}_n = \mathbb{O}, \forall n \in \mathbb{N}$. The utility function of player n is determined by $U1_n(O_n, O_{-n})$, where $O_n \in A_n$ is the chosen action of player n and O_{-n} is the action profile of all the active players except n. For an active player $n \in \mathbb{B}$, we define the utility function as follows:

$$U1_n(\mathbb{B}, O_n, O_{-n}) = - \sum_{m \in \mathbb{B}, m \neq n} O_n \odot O_m, \qquad (5.8)$$

which means that the state-based channel sensing order selection game can be expressed as follows:

$$G_1: \quad \max \; U1_n(\mathbb{B}, O_n, O_{-n}), \forall n \in \mathbb{B} \qquad (5.9)$$

5.3.1.2 Analysis of the Nash Equilibrium

Following the definition of pure strategy Nash equilibrium (NE), a channel sensing order profile $a_{NE} = \{O_n^*, O_{-n}^*\}$ is a pure strategy NE of $\mathbb{G}_1$ if and only if no player can improve its utility function by deviating unilaterally, i.e.,

$$U1_n(\mathbb{B}, O_n^*, O_{-n}^*) \geq U1_n(\mathbb{B}, O_n, O_{-n}^*), \forall O_n \in \mathbb{O}, \forall n \in \mathbb{B} \qquad (5.10)$$

In addition, the aggregate interference level in a pure strategy NE of any active user set $\mathbb{B}$ is given by

$$I_{\mathbb{G}_1} = -\sum_{n \in \mathbb{B}} U1_n(\mathbb{B}, O_n^*, O_{-n}^*) \qquad (5.11)$$

Theorem 5.1 *For any active user set, i.e., $\forall B \in \Gamma$, the state-based channel sensing order selection game G_1 is an exact potential game which has at least one pure strategy NE point. Furthermore, any optimal solution of problem $P1$ constitutes a pure strategy NE of the game.*

Proof According to the definition of exact potential games (see Definition 1.2 in Chap. 1), we need to prove that there exists a potential function such that the change in the utility function of an active player by its unilaterally deviating is the same as that in the potential function. To achieve this, a state-based potential function $\Phi1 : O_n \times O_{-n} \rightarrow R$ is defined as follows:

$$\Phi1(\mathbb{B}, O_n, O_{-n}) = -\frac{1}{2}\sum_{n \in \mathbb{B}}\sum_{k \in \mathbb{B}, k \neq n} O_n \odot O_m, \qquad (5.12)$$

which is exactly the negative half value of the aggregate interference level of all the active users. (Detailed proof lines are not presented here and can be found in [15].) Then, it can be verified that it is a potential game. Furthermore, using the relationship between the aggregate interference level and the potential function, as specified by (5.5) and (5.12), respectively, we can see that any optimal solution of problem **P1** is a global maximizer of the potential function. Then, following the property of exact potential games, Theorem 5.1 follows. □

Theorem 5.2 *For any active user set, i.e., $\forall \mathbb{B} \in \Gamma$, an interference-free channel sensing order profile always exists.*

Proof We prove this theorem by the method of construction. Without loss of generality, denote $B_1 = \{b_{11}, b_{12}, \ldots, b_{1M}\}$ as a channel sensing order vector arbitrarily chosen from the permutation set of M, i.e., $B_1 \in \mathbb{O}$. Construct B_2 by a cyclic shift of B_1, i.e., $B_2 = \{b_{12}, b_{13}, \ldots, b_{1M}, b_{11}\}$, and iteratively B_k by a cyclic shift of B_{k-1}, $k = 3, 4, \ldots, M$. Then, $B_k, k = 1, \ldots, M$, form the following matrix:

$$B_{cs} = \begin{bmatrix} b_{11} & b_{12} & \cdots & b_{1(M-1)} & b_{1M} \\ b_{12} & b_{13} & \cdots & b_{1M} & b_{11} \\ \vdots & \vdots & \vdots & \vdots & \vdots \\ b_{1(M-1)} & b_{1M} & \cdots & b_{1(M-3)} & b_{1(M-2)} \\ b_{1M} & b_{11} & \cdots & b_{1(M-2)} & b_{1(M-1)} \end{bmatrix}. \tag{5.13}$$

which is called the *cyclic-shift matrix*. It can be verified that the following equation always holds:

$$B_i \odot B_j = 0, \forall i, j \in \{1, 2, \ldots, M\}, i \neq j \tag{5.14}$$

Consequently, it follows that

$$\sum_{i=1}^{M} \sum_{j=1, j\neq i}^{M} B_i \odot B_j = 0. \tag{5.15}$$

Thus, it is known that $\{B_1, B_2, \ldots, B_M\}$ constitutes an interference-free profile for the full user set $\mathbb{N}$. For an arbitrary active user set $\mathbb{B}$, $|\mathbb{B}|$ distinct order vectors among B_{cs} can be chosen, which is also interference-free. Therefore, Theorem 5.2 is proved. □

If there are M channels, then the total number of cyclic-shift matrices is given the permutation number of $M - 1$, i.e., $(M - 1)!$. For example, there are total six cyclic-shift matrices of interference-free profiles for $M = 4$, which are shown in Fig. 5.2. Based on Theorem 5.2, the achievable performance of the state-based game can be studied.

Theorem 5.3 *For any active user set, i.e., $\forall \mathbb{B} \in \Gamma$, the best pure strategy NE of G_1 is an interference-free sensing order profile.*

Proof For $\forall \mathbb{B} \in \Gamma$, an interference-free channel sensing order profile always exists. The most efficient approach is to allocate different rows of a common cyclic-shift matrix, as characterized by (5.13), to the active users. Clearly, the optimal solution to problem $P1$ is an interference-free channel sensing order profile. Thus, according to Theorem 5.1, Theorem 5.3 follows. □

Fig. 5.2 For $M = 4$, there are six cyclic-shift matrices which correspond to interference-free order selection profiles

$$\begin{bmatrix} 1 & 2 & 3 & 4 \\ 2 & 3 & 4 & 1 \\ 3 & 4 & 1 & 2 \\ 4 & 1 & 2 & 3 \end{bmatrix} \quad \begin{bmatrix} 1 & 3 & 2 & 4 \\ 3 & 2 & 4 & 1 \\ 2 & 4 & 1 & 3 \\ 4 & 1 & 3 & 2 \end{bmatrix} \quad \begin{bmatrix} 1 & 4 & 2 & 3 \\ 4 & 2 & 3 & 1 \\ 2 & 3 & 1 & 4 \\ 3 & 1 & 4 & 2 \end{bmatrix}$$

$$\begin{bmatrix} 1 & 2 & 4 & 3 \\ 2 & 4 & 3 & 1 \\ 4 & 3 & 1 & 2 \\ 3 & 1 & 2 & 4 \end{bmatrix} \quad \begin{bmatrix} 1 & 3 & 4 & 2 \\ 3 & 4 & 2 & 1 \\ 4 & 2 & 1 & 3 \\ 2 & 1 & 3 & 4 \end{bmatrix} \quad \begin{bmatrix} 1 & 4 & 3 & 2 \\ 4 & 3 & 2 & 1 \\ 3 & 2 & 1 & 4 \\ 2 & 1 & 4 & 3 \end{bmatrix}$$

The above analysis results are promising since the best pure strategy NE of the G_1 corresponds to an optimal solution of **P1**. However, the state-based game cannot be solved as the system state $\mathbb{B}$ is random and unknown to the players. Thus, we develop a new game model for **P2** in the following.

5.3.2 Robust Order Selection Game

In this part, a robust order selection game, which can be regarded as an expected version of the state-based order selection game over all system states, is formulated to address the optimization problem **P2**. Formally, the robust order selection game is denoted as $\mathbb{G}_2 = \{\mathbb{N}, \{A_n\}_{n\in\mathbb{N}}, \{U2_n\}_{n\in\mathbb{N}}\}$, where $\mathbb{N}$ is the player set, A_n is the action set, and $U2_n$ is the utility function of player n. Note that the players in G_2 are all the potential users. The utility function is defined as follows:

$$U2_n(\mathbf{O}_n, \mathbf{O}_{-n}) = \mathrm{E}_{\mathbb{B}}\big[U1_n(\mathbb{B}, \mathbf{O}_n, \mathbf{O}_{-n})\big] = \sum_{\mathbb{B}\in\Gamma} \mu(\mathbb{B})U1_n(\mathbb{B}, \mathbf{O}_n, \mathbf{O}_{-n}), \qquad (5.16)$$

where $\mathrm{E}_{\mathbb{B}}$ takes the expectation over all system states. Similarly, the robust channel sensing order selection game can be expressed as

$$G_2: \qquad \max\ U2_n(\mathbf{O}_n, \mathbf{O}_{-n}), \forall n \in \mathbb{N} \qquad\qquad (5.17)$$

Based on the theoretic analysis for the state-based game, the properties of the NE of the robust order selection game are characterized by the following theorems.

Theorem 5.4 *The robust order selection game G_2 is also an exact potential game which has at least one pure strategy NE point. More importantly, any optimal solution of problem* **P2** *constitutes a pure strategy NE of G_2.*

Proof Refer to [15]. □

Theorem 5.5 *The best pure strategy NE of G_2 corresponds to an interference-free sensing order profile.*

Proof Based on Theorem 5.2 and following the similar lines for the proof of Theorem 5.3, this statement can be proved.

Finally, the robust order selection game G_2 and the state-based game G_1 are related by the following theorem.

Theorem 5.6 *A best pure strategy NE of the robust order selection game G_2 is also a best pure strategy NE of the state-based game G_1 for all system states, i.e., $\forall\mathbb{B} \in \Gamma$.*

Proof According to the above analysis, it is known that any best pure strategy NE of G_2 is interference-free. For any arbitrary active user set $\mathbb{B}$ in the state-based game

G_1, let the active users choose the actions drawn from a best pure strategy NE of G_2. As a result, the chosen channel sensing order vector profile is interference-free, which corresponds to a pure strategy NE of G_1. Therefore, Theorem 5.6 is proved.

As analyzed before, due to the restriction that no information exchange among the users is available and the active probabilities are unknown, it is not feasible to use traditional approaches to solve the problems **P1** and **P2**. In the following, a distributed learning algorithm that asymptotically converges to the best pure strategy NE solutions is proposed.

5.3.3 Distributed Learning Algorithm with Dynamic Active User Set

There are a large number of learning algorithms converging to the NE of potential games, e.g., best (better) response [16], spatial adaptive play [17], log-linear learning algorithms [18, 19] in Chaps. 3 and 4, and no-regret learning [20, 21]. However, these algorithms cannot be applied in the considered network with dynamic active user set.

Algorithm 4: *stochastic learning algorithm with dynamic active user set*

Initialization: at the first time being active, each active user n sets its channel selection as $\mathbf{q}_n(k_0) = (\frac{1}{M}, \ldots, \frac{1}{M})$.
Loop for $k = 0, 1, 2, \ldots$,
1. Selecting channel sensing orders: at the beginning of slot k, each active player $n \in \mathbb{B}(k)$ selects a channel sensing order $a_n(k) \in A_n$ according to its current mixed strategy $\mathbf{q}_n(k)$.
2. Sensing channels and receiving binary feedback: All the active players sense the licensed channels sequentially according to its chosen order and then transmit in the first idle channel. At the end of slot k, each active player receives a binary feedback $r_n(k)$, which is jointly determined by the channel states and the actions of other active users.
3. Updating mixed strategy: All the active users update their mixed strategies using the following rule:

$$\mathbf{q}_n(k+1) = \mathbf{q}_n(k) + br_n(k)\big(\mathbf{I}_{a_n(k)} - \mathbf{q}_n(k)\big), \qquad (5.18)$$

where $0 < b < 1$ is the learning parameter, $\mathbf{I}_{a_n(k)}$ is a unit vector with the $a_n(k)$th component being one. All inactive users keeps their mixed strategies unchanged.
End loop

In this chapter, a stochastic learning automata [22]-based algorithm is proposed for the robust order selection game. The players employ mixed strategies in each slot. Specifically, denote $\mathbf{Q}(k) = (\mathbf{q}_1(k), \ldots, \mathbf{q}_n(k))$ as the mixed strategy profile in slot k, in which $\mathbf{q}_n(k) = (q_{n1}(k), \ldots, q_{nM}(k))$ is the action probability vector of player n. The algorithm iterates as follows: (i) in the first slot player n being active, it randomly chooses a channel sensing order with equal probabilities, i.e.,

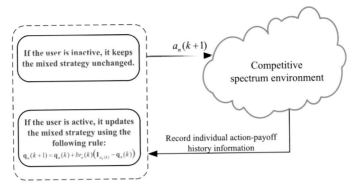

Fig. 5.3 The illustrative diagram of the stochastic automata-based learning algorithm with dynamic active user set

$\mathbf{q}_n(k_0) = (\frac{1}{M}, \ldots, \frac{1}{M})$, (ii) for an active player $n \in \mathbb{B}(k)$ in slot k, it then employs an rule to update its mixed strategy based on the received binary feedback $r_n(k)$. An inactive user is always silent and keeps its mixed strategy unchanged. The proposed stochastic learning algorithm with dynamic active user set is formally described in Algorithm 4 and an illustrative diagram is shown in Fig. 5.3.

The learning algorithm is further discussed below. For presentation, the slot index is added into the game models. Specifically, denote $s_n(k)$ as the state of player n in slot k, $\mathbb{B}(k)$ as the active user set in slot k, and $r_n(k)$ is the received binary feedback of player n. Based on the transmission strategies of the users, the binary feedback is determined as follows: (i) $r_n(k) = 1$, which indicates that the transmission of player n is successful, and (ii) $r_n(k) = 0$, which indicates that player n experiences a collision or does find idle channels. As the action space is extremely huge, e.g., a number of the action space for $M = 5, 6$ are 120 and 720, respectively, it is desirable to reduce the action space size for accelerating the convergence speed. A simple and efficient approach is used as follows: all the active users use a common cyclic-shift matrix as their action space, i.e., $A_n = \mathbf{B}_{cs}, \forall n \in \mathbb{N}$.

Theorem 5.7 *When a common cyclic-shift matrix is used as the action space for all the players, the proposed stochastic learning algorithm asymptotically converges to the best pure strategy NE points of both G_1 and G_2 if the learning parameter goes sufficiently small, i.e., $b \to 0$.*

Proof The detail proof lines are not presented here but can be found in [15]. Following the methodology for the convergence of the stochastic learning automata algorithm presented in Chap. 2, and with some modification in dealing with the dynamic active user set, similar lines can be used to prove the convergence. Moreover, since all the players use a common cyclic-shift matrix as the action set, it is concluded that the proposed learning algorithm converges to the best Nash equilibria of G_1 and G_2, according to Theorems 5.3, 5.5, and 5.6. Therefore, Theorem 5.7 can be proved.   □

5.4 Simulation Results and Discussion

To begin with, the parameters and scenario setting for simulation are specified. The normalized achievable throughput of an active player successfully accessing at the nth channel in a slot is given by $R = 1 - n\tau$, where $\tau = \frac{T_s}{T}$ is the sensing time fraction in a slot. Following the similar setting in [23], the slot length is set to $T = 100$ ms and the sensing duration is set to $T_s = 5$ ms. For its simplicity and easy implementation, the energy detection approach [23] is employed and the spectrum sensing performance is characterized by $P_d(T_s) = 0.9$ and $P_f(T_s) = 0.1$. Furthermore, for convenience of discussion, it is assumed that the idle probabilities of the channels are the same, i.e., $\theta_m = \theta, \forall m \in \{1, 2, \ldots, M\}$, and the active probabilities of the users are also the same, i.e., $\lambda_n = \lambda, \forall n \in \{1, 2, \ldots, N\}$, otherwise specified.

5.4.1 Convergence Behavior

In this part, the convergence behaviors of the proposed learning algorithm in the presence of dynamic active users are studied. Consider a network with five channels and five users, i.e., $M = 5, N = 5$. Assume that the channel idle probabilities are $\theta = 0.6$ and the user active probabilities are $\lambda = 0.5$. The step size in the learning algorithm is set to $b = 0.05$, which has been optimized by experiment.

First, for an arbitrarily chosen user, the evolution of its order selection probabilities is shown in Fig. 5.4. It can be observed that the probabilities remain unchanged in successive multiple slots, e.g., from slot 200 to 215, which is caused by the event that it is inactive in these slots. The mixed strategy finally converges to a pure strategy action ($\mathbf{q} = \{0, 0, 1, 0, 0\}$) in about 400 iterations (slots). After slot 400, the player adheres to the converging stable solution. Furthermore, the evolution of the

Fig. 5.4 The evolution of the order selection probabilities of an arbitrarily chosen users ($M = 5$, $N = 5, \theta = 0.6, \lambda = 0.5$)

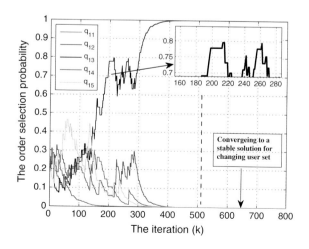

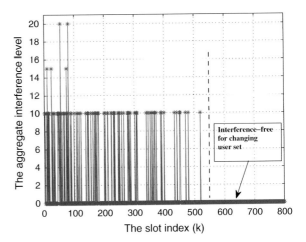

Fig. 5.5 The evolution of the aggregate interference level (the number of channels is $M = 5$, the number of users is $N = 5$, the channel idle probabilities are $\theta_n = 0.6$, and the active probabilities of the users are $\lambda = 0.5$)

aggregate interference level is shown in Fig. 5.5. It is noted from the figure that it finally decreases to zero in about 500 iterations, which implies that the converging order selection profile is interference-free for every changing active user set. Thus, the results presented in the figures validate the convergence and optimality of the proposed learning algorithm in the presence of dynamic active user set.

Second, as the proposed learning algorithm takes several slots to converge to stable solutions, its convergence time should be studied. Specifically, the expected convergence time is defined as the number of iterations that when a component of the mixed strategy of each player is sufficiently approaching one, e.g., larger than 0.95. The expected convergence times when varying the number of users and the active probabilities are shown in Fig. 5.6. Some important observations are as follows:

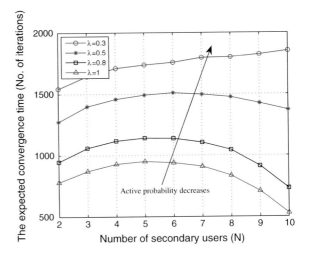

Fig. 5.6 The expected convergence time (No. of iterations) with different numbers of users and active probabilities in each slot (the number of channels is $M = 10$, and the channel idle probabilities are $\theta_n = 0.8$)

- For a specific number of users, e.g., $N = 5$, the expected convergence time increases as the user active probability decreases. The reason is that a user with lower active probability performs learning occasionally, while the one with higher active probability does more frequently.
- For scenarios with large active probabilities, e.g., $\lambda = 0.5, 0.8, 1$, the expected convergence time increases when the number of users is small, e.g., $N < 6$, and decreases when the number of users becomes large, e.g., $N \geq 6$.

5.4.2 Throughput Performance

In this part, the throughput performance of the proposed learning approach is studied. Specifically, we compare the throughput performance of the proposed learning approach with the random selection approach, in which each active player chooses the sensing and access order randomly and autonomously. To study the effect of channel availability on the system throughput, the following four scenarios are considered:

- (Scenario 1) homogeneous channels: all the channel idle probabilities are set to 0.8.
- (Scenario 2) slight heterogeneous channels: the channel idle probabilities are set to $\{0.7, 0.7, 0.7, 0.8, 0.8, 0.8, 0.8, 0.9, 0.9, 0.9\}$.
- (Scenario 3) moderate heterogeneous channels: the channel idle probabilities are set to $\{0.5, 0.5, 0.6, 0.6, 0.7, 0.7, 0.8, 0.8, 0.9, 0.9\}$.
- (Scenario 4) heavy heterogeneous channels: the channel idle probabilities are set to $\{0.1, 0.2, 0.3, 0.4, 0.5, 0.6, 0.7, 0.8, 0.8, 0.9\}$.

The comparison results versus the user active probabilities are shown in Fig. 5.7. The results are obtained by simulating 100,000 successive slots and then taking the

Fig. 5.7 The comparison results versus the active probabilities of the users in each slot (the number of channels is $M = 10$, and the number of users is $N = 10$)

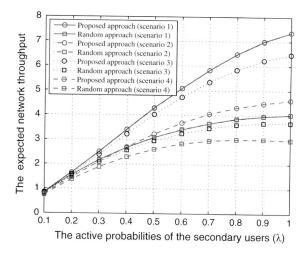

Fig. 5.8 The comparison results versus the number of users (the number of channels is $M = 10$, and the active probability of each user is $\lambda = 0.7$)

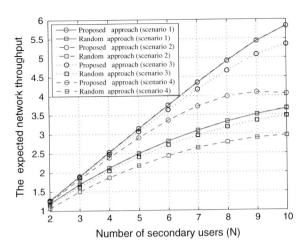

expected value. It can be observed that the normalized expected system throughput of both approaches increases as the user active probability λ increases. In addition, the comparison results versus the number of users are shown in Fig. 5.8. The active probability of each secondary user is set to $\lambda = 0.7$. It is noted that the normalized expected system throughput of both approaches increases as the number of users increases. In addition, the proposed learning algorithm outperforms the random selection approach for four considered scenarios significantly.

To summarize, the simulation results validate the convergence of the proposed learning algorithm in the presence of dynamic active user set. Also, it is shown that it outperforms the random approach significantly in both homogeneous and heterogeneous scenarios.

5.5 Concluding Remarks

Compared with the interference modes considered in Chaps. 2, 3, and 4, we can see that there are two new differences in this chapter: (i) the active user set is time-varying, which is more general in practical wireless networks; and (ii) a generalized interference metric is used to address the overlap in channel sensing order. Note that following the methodology in this chapter, some studies on resource optimization problems with dynamic active user set were recently reported in [24–26]. Thus, it is believed that the model and results presented in this chapter provide efficient solutions for multiuser resource optimization problems.

References

1. Q. Zhao, L. Tong, A. Swami et al., Decentralized cognitive MAC for opportunistic spectrum access in ad hoc networks: a POMDP framework. IEEE J. Sel. Areas Commun. **25**(3), 589–600 (2007)
2. Y. Xu, A. Anpalagan, Q. Wu et al., Decision-theoretic distributed channel selection for opportunistic spectrum access: strategies, challenges and solutions. IEEE Commun. Surv. Tutor. Fourth Quarter **15**(4), 1689–1713 (2013)
3. Z. Khan, J. Lehtomäki, L. DaSilva et al., Autonomous sensing order selection strategies exploiting channel access information. IEEE Trans. Mob. Comput. **12**(2), 274–288 (2013)
4. J. Zhao, X. Wang, Channel sensing order in multi-user cognitive radio networks, Proceedings of IEEE International Symposium on Dynamic Spectrum Access Networks. (IEEE Publications, Bellevue, 2012)
5. A. Mendes, C. Augusto, M. Silva et al., Channel sensing order for cognitive radio networks using reinforcement learning. 36th Annual IEEE Conference on Local Computer Networks. pp. 546–553 (2011)
6. H. Shokri-Ghadikolaei, F. Sheikholeslami, M. Nasiri-Kenari, Distributed multiuser sequential channel sensing schemes in multichannel cognitive radio networks. IEEE Trans. Wirel. Commun. **12**(5), 2055–2067 (2013)
7. S. Kim, G. Giannakis, Sequential and cooperative sensing for multi-channel cognitive radios. IEEE Trans. Signal Process. **58**(8), 4239–4253 (2010)
8. H. Jiang, L. Lai, R. Fan, H. Poor, Optimal selection of channel sensing order in cognitive radio. IEEE Trans. Wirel. Commun. **8**(1), 297–307 (2009)
9. N. Chang, M. Liu, Optimal channel probing and transmission scheduling for opportunistic spectrum access. IEEE/ACM Trans. Netw. **17**(6), 1805–1818 (2009)
10. Y. Xu, J. Wang, Q. Wu et al., Optimal energy-efficient channel exploration for opportunistic spectrum usage. IEEE Wirel. Commun. Lett. **1**(2), 77–80 (2012)
11. H. Cheng, W. Zhuang, Simple channel sensing order in cognitive radio networks. IEEE J. Sel. Areas Commun. **29**(4), 676–688 (2011)
12. T. Shu, H. Li, QoS-compliant sequential channel sensing for cognitive radios. IEEE J. Sel. Areas Commun. **32**(11), 2013–2025 (2014). (to appear)
13. Y. Pei, Y.-C. Liang, K. Teh et al., Energy-efficient design of sequential channel sensing in cognitive radio networks: optimal sensing strategy, power allocation, and sensing order. IEEE J. Sel. Areas Commun. **29**(4), 1648–1659 (2011)
14. H. Tembine, *Distributed Strategic Learning for Wireless Engineers* (CRC Press, Boca Raton, 2012)
15. Y. Xu, Q. Wu, L. Shen, J. Wang, A. Anpalgan, Robust multiuser sequential channel sensing and access in dynamic cognitive radio networks: potential games and stochastic learning. IEEE Trans. Veh. Technol. **64**(8), 3594–3607 (2015)
16. D. Monderer, L.S. Shapley, Potential games. Games Econ. Behav. **14**, 124–143 (1996)
17. Y. Xu, J. Wang, Q. Wu et al., Opportunistic spectrum access in cognitive radio networks: global optimization using local interaction games. IEEE J. Sel. Signal Process **6**(2), 180–194 (2012)
18. Y. Xu, Q. Wu, J. Wang et al., Opportunistic spectrum access using partially overlapping channels: graphical game and uncoupled learning. IEEE Trans. Commun. **61**(9), 3906–3918 (2013)
19. Y. Xu, Q. Wu, L. Shen et al., Opportunistic spectrum access with spatial reuse: graphical game and uncoupled learning solutions. IEEE Trans. Wirel. Commun. **12**(10), 4814–4826 (2013)
20. N. Nie, C. Comaniciu, Adaptive channel allocation spectrum etiquette for cognitive radio networks. Mobile Netw. Appl. **11**(6), 779–797 (2006)
21. M. Maskery, V. Krishnamurthy, Q. Zhao, Decentralized dynamic spectrum access for cognitive radios: cooperative design of a non-cooperative game. IEEE Trans. Commun. **57**(2), 459–469 (2009)
22. P. Sastry, V. Phansalkar, M. Thathachar, Decentralized learning of nash equilibria in multi-person stochastic games with incomplete information. IEEE Trans. Syst., Man, Cybern. B **24**(5), 769–777 (1994)

23. Y.-C. Liang, Y. Zeng, E. Peh et al., Sensing-throughput tradeoff for cognitive radio networks. IEEE Trans. Wirel. Commun. **7**(4), 1326–1337 (2008)
24. J. Zheng, Y. Cai, Y. Xu, A. Anpalagan, Distributed channel selection for interference mitigation in dynamic environment: a game-theoretic stochastic learning solution. IEEE Trans. Veh. Technol. **63**(9), 4757–4762 (2014)
25. J. Zheng, Y. Cai, N. Lu, Y. Xu, X. Sherman Shen, Stochastic game-theoretic spectrum access in distributed and dynamic environment. IEEE Trans. Veh. Technol. doi:10.1109/TVT.2014. 2366559
26. J. Zheng, Y. Cai, W. Yang, Y. Xu, A. Anpalgan, A game-theoretic approach to exploit partially overlapping channels in dynamic and distributed networks. IEEE Commun. Lett. **18**(12), 2201–2204 (2014)

Chapter 6
Future Direction and Research Issues

In this chapter, we briefly discuss some future direction and research issues with regard to interference mitigation in the fifth-generation mobile communication systems (5G).

6.1 Hierarchical Games for Small Cell Networks

With the ever-increasing demand for high-speed and high-quality wireless data applications, e.g., video streaming, online gaming, and social networks, small cell technology is emerging as a powerful and economic solution to boost the system capacity and enhance the network coverage [1, 2]. Generally, typical small cells include the operator-deployed micro-cells and pico-cells as well as the user-deployed femto-cells. Specifically, femtocells are low-power and short-range access points, which are mainly applied to improve the indoor experience of cellular mobile users and managed by end users in a plug-and-play manner. Because small base stations (SBSs) are deployed in the coverage range of a macro base station (MBS), from the perspective of network operators, SBSs can drastically improve the spectrum efficiency due to spatial reuse and offloading partial traffic load from the main network.

In practice, from the perspective of either an infrastructure or spectrum availability, it is more favorable to deploy two-tier small cell networks in shared spectrum rather than splitting spectrum scheme [3]. However, the co-existing issue of co-channel deployed SBSs and MBSs brings about numerous technical challenges in terms of interference management. Without proper interference control, the cross-tier and co-tier interferences severely affect the overall system performance. Accordingly, the interference mitigation is an important research area and is regarded as the major challenge in spectrum-sharing small cell networks.

Various interference mitigation schemes have been proposed for heterogeneous wireless networks [4–7]. However, these approaches cannot be directly applied to practical two-tier small cell networks, as these schemes are centralized and hence

© The Author(s) 2016
Y. Xu and A. Anpalagan, *Game-theoretic Interference Coordination*
Approaches for Dynamic Spectrum Access, SpringerBriefs in Electrical
and Computer Engineering, DOI 10.1007/978-981-10-0024-9_6

need coordination between SBSs and MBSs. As a result, it requires a large number of timely cross-tier and co-tier information exchange and leads to heavy overhead especially in large-cale scenarios. In addition, because of the randomness of mobile users' activity and the small cell access points' placement, it results in the ad-hoc topology of small cell networks, which implies that the networks' topology is essentially affected by end users' behavior. Therefore, centralized optimization approaches seem to be impractical, and hence it is desirable to develop distributed interference management approaches for small cell networks.

Due to the hierarchical decision structure between MBSs and SBSs, it is suitable and natural to apply the Stackelberg game, also known as leader–follower game or hierarchical game, to model the hierarchical interaction and competition between MBS and SBS in two-tier networks. Specifically, the MBS is modeled as leader and moves first. In the sequel, the SBSs are followers and take their actions based on the observation of leaders' actions. Note that the Stackelberg game is the extension of normal non-cooperative game. In a typical non-cooperative hierarchical game, the Stackelberg equilibrium (SE) is commonly used as a universal solution concept. SE is a stable operation point, at which no player can improve its utility by deviating unilaterally in the hierarchical game, which has the similar meaning as Nash equilibrium (NE) in formal game. The hierarchical game addresses the differentiated demands and priorities in tiered communication systems, and thus, it has been shown in [3, 8–13] that Stackelberg games provide a suitable framework to implement interference management in two-tier small cell networks.

Technically, when it comes to apply the Stackelberg games into small cell networks, the following concerns should be addressed:

- **Sophisticated utility function design**. In the hierarchical game, both the leaders and followers are aiming at maximizing their own utility function. Utility function reflects the differentiation demand and preference of player involved in the game. To avoid the tragedy of commons, leading to inefficient SE, the utility function should be well designed to achieve improved performance in concerned metrics. Moreover, the utility function should have the specific physical meaning such as achievable rate, quality of experience (QoE), and energy efficiency.
  To obtain the SE in a relative simple and low-computation way, the utility functions are commonly well designed to satisfy some features, i.e., the utility functions are designed as concave function shown in [3, 9] to facilitate the usage of backward induction, so that the SE can be obtained in a closed form. On the other hand, getting the closed-form SE may be not an easy task, and some works are expecting to obtain SE in a recursive manner that requires the utility function to meet some specific characteristics [10]. For example, if the players' strategy updating strategy follows the standard interference function first introduced by Yates [14], it can be guaranteed to converge to the stable operation point which admits an SE.
- **Robust decision under information uncertainty**. In practical systems, the assisted information for decision making may contain uncertainty, i.e., some parameters involved in the utility function or constraints cannot be precisely observed. The uncertainty may be caused by dynamic communication environment such

as time-varying channel state information (CSI), information transmission error caused by limited feedback bandwidth, and so on. The imperfect information scenario is more practical than perfect assumption applied in [9, 11–13]. However, if not well designed and optimized, the performance may degrade drastically under the SE which is obtained with the perfect information assumption. From the perspective of optimization theory, there are two widely used approaches to deal with the information uncertainties in game models [11]: (i) Bayesian approach [9]: it considers the average payoff based on some prior distribution information, and (ii) Robust optimization [11]: it considers the payoff for the worst-case scenario, which is a distribution-free method.

Another efficient method is resorting to learning theory, e.g., the stochastic learning automata [15]. Under the hierarchical learning framework, the SBS and MBS are assumed to behave as intelligent agents and have self-learning ability to automatically optimize their configuration. Each smart agent's overall goal is to learn to optimize its individual long-term cumulative reward via repeatedly interacting with network environment [16].

- **Scalability with dense deployment**. Network densification is the dominated theme in the 5G communication systems [1]. However, the problem of interference management under hyper-dense deployment of SBS is still an open issue. The cost of obtaining the SE solution in large-scale scenario may drastically increase. One possible approach is to split the large-scale optimization problem into several dub-problems. Another possible solution is to build a smart decision system based on cloud infrastructure owning powerful real-time computational capabilities. Using the data mining technique and with the assistance of information base in the cloud center, the users in system can utilize the mixed information, e.g., feedback knowledge from either learning or reasoning, and multi-dimensional context information including spectrum state, channel state information, location, energy, etc., to make efficient decisions. In addition, the users can utilize machine learning methods, e.g., online learning and statistical learning in dynamic scenarios, to make decisions more flexible, efficient, and intelligent.

6.2 Interference Mitigation for Carrier Aggregation

Carrier aggregation (CA) [17] has been regarded as a promising technology for 5G systems, as multiple spectrum bands can be simultaneously utilized to satisfy the large bandwidth demand. Compared with most existing interference mitigation approaches, there are some new challenges and problems for interference mitigation with carrier aggregation. In particular, the inherent characteristics of CA should be well addressed. However, this problem just begins to draw attentions very recently and some research issues are listed below.

- **The cost of non-contiguous CA should be addressed**. In general, CA can be used in three different scenarios: intra-band contiguous CA, intra-band non-contiguous

CA, and inter-band non-contiguous CA. For intra-band contiguous CA, it needs only one fast Fourier transform (FFT). For intra-band non-contiguous CA, it is more complicated than the intra-band contiguous CA. Specifically, the multi-carrier signal cannot be treated as a single signal and hence more transceivers are required, which adds processing complexity significantly. For inter-band non-contiguous CA, it requires separate FFT and needs the use of multiple transceivers. In addition, reducing inter-modulation and cross-modulation from different transceivers also causes future complexity. Thus, the cost for non-contiguous CA should be included in the interference mitigation problem [18], which differs significantly from existing interference mitigation approaches.

- **Autonomous guard band assignment**. In practice, guard bands are needed to prevent mutual interference among multiple users. The guard bands naturally constrain the spectrum utilization. Furthermore, fixed guard bands are not suitable for scenarios where multiple operators independently and autonomously perform CA. Thus, it is timely and important to achieve autonomous guard band assignment. In [19], the authors investigated the problem of assigning channels/powers to opportunistic transmissions, taking into account the constraint of guard bands, and then proposed a guard-band-aware channel assignment scheme for dynamic spectrum access systems with CA. Based on the above preliminary results, further investigation is needed.
- **The heterogeneous channel availabilities cause a new interference paradigm**. In practice, the licensed users always transmit at different times and on different channels, i.e., the channel availabilities are heterogeneous. From the view of interference coordination, the transmission of unlicensed users is dramatically affected if one of the aggregated channels is subjected to severe interference, even when the other aggregated channels are interference-free. This interference diagram is different from traditional interference models without CA, e.g., the modes presented in Chaps. 2–5. Thus, how to model and analyze the new interference paradigm for CA is important and urgent.

References

1. J.G. Andrews, H. Claussen, M. Dohler et al., Femtocells: past, present, and future. IEEE J. Sel. Areas Commun. **30**(3), 497–508 (2012)
2. J.G. Andrews, S. Buzzi, W. Choi et al., What will 5G be? IEEE J. Sel. Areas Commun. **32**(6), 1065–1082 (2014)
3. X. Kang, R. Zhang, M. Motani, Price-based resource allocation for spectrum-sharing femtocell networks: a Stackelberg game approach. IEEE J. Sel. Areas Commun. **30**(3), 538–549 (2012)
4. V. Chandrasekhar, J.G. Andrews, T. Muharemovic, Z. Shen, Power control in two-tier femtocell networks. IEEE Trans. Wireless Commun. **8**(8), 4316–4328 (2009)
5. H. Zhang, C. Jiang, N.C. Beaulieu et al., Resource allocation in spectrum-sharing OFDMA femtocells with heterogeneous services. IEEE Trans. Commun. **62**(7), 2366–2377 (2014)
6. D. Nguyen, T. Le-Ngoc, Sum-rate maximization in the multi cell MIMO multiple-access channel with interference coordination. IEEE Trans. Wirel. Commun. **13**(1), 36–48 (2014)

7. C. Tan, S. Friedland, S. Low, Non-negative matrix inequalities and their application to non-convex power control optimization. SIAM J. Matrix Anal. Appl. **32**(3), 1030–1055 (2011)
8. S. Guruacharya, D. Niyato, D. Kim, E. Hossain, Hierarchical competition for downlink power allocation in OFDMA femtocell networks. IEEE Trans. Wirel. Commun. **12**(4), 1543–1553 (2013)
9. S. Bu, F. Yu, Y. Cai, H. Yanikomeroglu, Interference-aware energy-efficient resource allocation for OFDMA-based heterogeneous networks with incomplete channel state information. IEEE Trans. Veh. Technol. **64**(3), 1036–1050 (2014)
10. Q. Han, B. Yang, X. Wang et al., Hierarchical-game-based uplink power control in femtocell networks. IEEE Trans. Veh. Technol. **63**(6), 2819–2835 (2014)
11. K. Zhu, E. Hossain, A. Anpalagan, Downlink power control in two-tier cellular OFDMA networks under uncertainties: a robust Stackelberg game, IEEE Trans. Commun. (to appear)
12. S. Parsaeefard, A. Sharafat, M. Schaar, Robust additively coupled games in the presence of bounded uncertainty in communication networks. IEEE Trans. Veh. Technol. **63**(3), 1436–1451 (2014)
13. S. Parsaeefard, M. Schaar, A. Sharafat, Robust power control for heterogeneous users in shared unlicensed bands. IEEE Trans. Wirel. Commun. **13**(6), 3167–3182 (2014)
14. R. Yates, A framework for uplink power cellular radio systems. IEEE J. Sel. Areas Commun. **13**(7), 1341–1347 (1995)
15. P. Sastry, V. Phansalkar, M. Thathachar, Decentralized learning of nash equilibria in multi-person stochastic games with incomplete information. IEEE Trans. Syst. Man Cybern. **24**(5), 769–777 (1994)
16. X. Chen, H. Zhang, T. Chen, et al., Improving energy efficiency in femtocell networks: a hierarchical reinforcement learning framework, in *Proceeding of the IEEE International Conference on Communications (ICC)*, Budapest, pp. 2241–2245, (2013)
17. Z. Khan, H. Ahmadi, E. Hossain, M. Coupechoux, Carrier aggregation channel/bonding in next generation cellular networks: methods and challenges. IEEE Netw. **28**(6), 34–40 (2014)
18. H. Ahmadi, I. Macaluso, L. DaSilva, Carrier aggregation as a repeated game: learning algorithms for convergence to a Nash equilibrium, in *IEEE Global Communications Conference (GLOBECOM)* (2013)
19. H. Salameh, M. Krunz, D. Manzi, Spectrum bonding and aggregation with guard-band awareness in cognitive radio networks. IEEE Trans. Mobile Comput. **13**(3), 569–581 (2014)

Printed in the United States
By Bookmasters